CAMBRIDGE MONOGRAPHS IN
EXPERIMENTAL BIOLOGY
No. 9

EDITORS:
M. ABERCROMBIE, P. B. MEDAWAR,
GEORGE SALT (*General Editor*)
M. M. SWANN, V. B. WIGGLESWORTH

INSECT FLIGHT

THE SERIES

INSECT FLIGHT

BY

J. W. S. PRINGLE
M.B.E., Sc.D., F.R.S.

*Fellow of Peterhouse, Lecturer in Zoology in the
University of Cambridge*

CAMBRIDGE
AT THE UNIVERSITY PRESS
1957

CAMBRIDGE UNIVERSITY PRESS
Cambridge, New York, Melbourne, Madrid, Cape Town, Singapore,
São Paulo, Delhi, Dubai, Tokyo

Cambridge University Press
The Edinburgh Building, Cambridge CB2 8RU, UK

Published in the United States of America by Cambridge University Press, New York

www.cambridge.org
Information on this title: www.cambridge.org/9780521135009

First published 1957
This digitally printed version 2010

A catalogue record for this publication is available from the British Library

ISBN 978-0-521-05995-4 Hardback
ISBN 978-0-521-13500-9 Paperback

CONTENTS

And under the firmament were their wings straight, the one toward the other: every one had two, which covered on this side, and every one had two, which covered on that side, their bodies.

And when they went, I heard the noise of their wings, like the noise of great waters, as the voice of the Almighty, the voice of speech, as the noise of an host: when they stood, they let down their wings.

EZEKIEL I. 23, 24.

PREFACE

THE flight of insects has interested zoologists from the earliest times. It is the most obvious of the characteristics which have made this class of animal successful in the colonization of a wide variety of terrestrial environments, and the insect's precision in aerial manoeuvres commands the admiration of the naturalist even in these days when man's conquest of the air is no longer a novelty. To the serious student, however, small size and the speed with which the wings are moved in the more efficient fliers combine to make accurate description of the mechanics of flight a more difficult task than for some of the other types of animal locomotion. It was only with the introduction of photographic and other instrumental aids towards the end of the nineteenth century that a beginning was made with the accumulation of quantitative data which could form the basis on the one hand for aerodynamic theories and on the other for physiological analysis of the properties of muscles and other structures comprising the flight system.

A landmark in the history of the subject was the publication in 1934 of Magnan's book, *Le Vol des Insectes*. The researches summarized in that book have provided the material for the chapters on insect flight in most text-books of entomology in the succeeding twenty years, and are still of great value. But it is impossible to read some of these accounts without being impressed by the apparent isolation of zoological thought of the time from the parallel progress being made in the sciences of aeronautics and comparative physiology. Reports about the speed of flight of certain flies were widely quoted which bore no relation to physical possibility, and theories were propounded in conflict with principles which were fundamental in other fields of study. One of the characteristics of modern biology is the breakdown of the boundaries which separate its subdivisions, and nowhere is this more fruitful than in behavioural studies where mechanics and physiology are integral to a more profound understanding. The new approach is well illustrated by L. E. Chadwick's chapters on insect flight in K. D. Roeder's

Insect Physiology, but progress has been sufficiently rapid since 1953 to justify a fresh review of our state of knowledge.

I wish particularly to acknowledge my indebtedness to Dr Torkel Weis-Fogh, who has been a constant source of help and inspiration in the preparation of this book, and who has himself made many of the discoveries which make possible a new synthesis. Many others have helped by discussion and in the search through the literature; especially I should like to thank Prof. E. G. Boettiger, Dr R. M. Baranowski, Dr Jean Hanson, and Dr J. Smart for allowing me to read and quote from their unpublished papers. Dr G. Salt kindly read the manuscript and made several useful comments.

J. W. S. P.

CAMBRIDGE
March 1957

CHAPTER I

General Anatomy of the Wings and of a Wing-bearing Segment

THIS monograph deals with the subject of insect flight from a functional and physiological point of view. In order to understand how an insect flies, however, it is necessary to have a clear picture of the anatomy, and in this introductory chapter a brief summary is given of the general plan of the skeletal and muscular structures comprising the machinery of aerial locomotion in the group. Detailed treatment of this aspect of insect flight is to be found in most text-books of entomology, in particular in Snodgrass (1935). Great differences of detail are found in the different orders, but a combination of anatomical and physiological studies has now made it possible to establish the essential features of the evolution of most of the main types of flight mechanism found in the Pterygota.

The wings of insects arose in the Devonian or Lower Carboniferous as lateral expansions of the thoracic nota, and it is generally agreed that their original function was to enable the insect to glide from trees to the ground; fossil types are known with pronotal as well as meso- and metanotal expansions of this region of the exoskeleton. This originally continuous fold between the tergal and pleural sclerotizations soon narrowed, and in all modern winged insects definite regions of wing articulation are formed by more elaborate folding and local thickening of the basal region. The pattern of tergal and pleural sclerotization in the meso- and metathorax is also constant in its main features (fig. 1). Both tergum and pleuron are divided into anterior and posterior portions by an internal fold. The scutum bears on its lateral edge the anterior notal process, which forms the main dorsal support for the wing; the posterior notal process is also formed by a fold of the scutum, but usually is rigid with the scutellum. The two main pleural sclerites, the episternum and epimeron, are separated by an internal ridge or suture, which serves to strengthen the lateral wall of the thorax and is often continued in its middle region into an apodeme for the attachment of the pleurosternal muscle. At its upper end the pleural ridge reaches the pleural wing process; below, it helps to form the main articulation for the coxa of the leg. The thoracic box is completed below by the sternum.

In the dorsal region of a wing-bearing segment there are usually present large, intersegmental phragmata which serve for the attachment of the dorsal longitudinal muscles. The phragmata are internal folds not of the tergum proper but of intercalated sclerites; the posterior intercalated sclerite, or

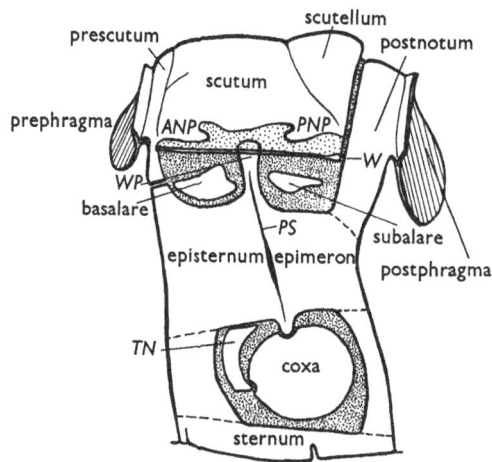

Fig. 1. Diagrammatic lateral view of a typical wing-bearing segment. *ANP*, anterior notal process; *PNP*, posterior notal process; *PS*, pleural suture; *TN*, trochantin; *W*, wing (cut through); *WP*, pleural wing process. (Redrawn from Snodgrass, 1935.)

postnotum, may be large and then forms a complete dorsal bridge between the two pleura behind the wing. A narrow anterior bridge may be formed between the prescutum and the episternum.

At the upper edge of the pleuron, below the wing fold, two regions, known as the basalare and subalare, may become detached as separate sclerites. These bear the dorsal insertion of important muscles which serve to twist the wing about the transverse body axis by ligamentous cuticular attachments to sclerites of the wing base.

The sclerotization of the wing itself is a subject which has been much studied by systematic entomologists, since the pattern of wing venation is often the best preserved feature of insect fossils, and serves as a ready means of identification of many orders and families. In a discussion of the origin of wings and of venational types, Forbes (1943) distinguishes four or five main evolutionary lines; of these, there is general agreement only that the Odonata and Ephemeroptera stand distinct from other modern orders in the structure of the basal wing sclerites and in their inability to fold the wings back over the abdomen when at rest. In the other

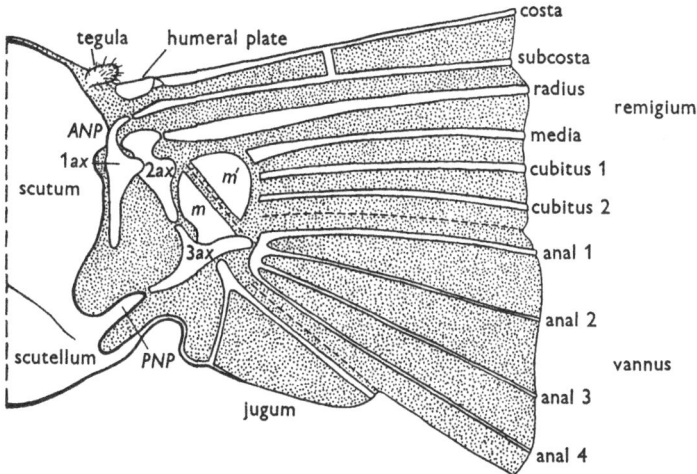

Fig. 2. Diagram showing a typical arrangement of the basal wing sclerites, and the naming of the principal wing areas and veins; the interrupted lines show the main regions of wing-folding. *ANP*, anterior notal process; *m, m'*, median plates; *PNP*, posterior notal process; *1ax, 2ax, 3ax*, axillary sclerites. (Redrawn, with alterations, from Snodgrass, 1935.)

orders of the Pterygota (Snodgrass, 1909, 1927; Comstock and Needham, 1898, 1899; Comstock, 1918), the wing veins, the basal sclerites and the wing muscles have an arrangement which can be referred to a basic plan, with a knowledge of which the various modifications are more readily visualized.

The three or four axillary sclerites bear a constant relationship to the wing processes of the notum and pleuron and to the main wing veins (fig. 2). The 1st axillary articulates with the anterior notal process, and in some orders (e.g. Hymenoptera) its posterior arm, which may be directed diagonally downwards, is also in close association with the posterior notal process.

Laterally the 1st axillary abuts on the end of the sub-costal vein and forms a complicated articulation with the 2nd axillary. The 2nd axillary usually supports the radius, and its lower surface rests on the pleural wing process; it may also have a strong connexion to the median plate. The 3rd axillary is the pivotal sclerite for the wing-flexing mechanism. It has three articulations: basally with the posterior notal process and the posterior arm of the 2nd axillary, and distally with the anal wing veins. A 4th axillary is present as a distinct sclerite only in Orthoptera and Hymenoptera and lies between the 3rd axillary and posterior notal process.

The power for the up- and downstrokes may have been derived primitively from the direct muscles; these are dominant in the Blattoidea and remain important in all Orthoptera, Coleoptera and Odonata. Elsewhere there is a tendency for the main power for the stroke to be provided by indirect muscles, which in some of the highest orders become very large and occupy most of the volume of one or both of the pterothoracic segments; the direct muscles here are reduced in size but always remain of importance for the control of the wing beat.

The indirect muscles are arranged in two functional groups. The dorsal longitudinal muscles, running from prephragma to postphragma (fig. 3A), are usually described as producing in the main an arching of the tergum, raising the notal processes relative to the pleural process and so depressing the wing (Chabrier, 1822). In fact, in many insects there is an equally important approximation of the anterior and posterior notal processes (Janet, 1899) due to the development on each side of the notum of a line of weakness between scutum and scutellum. The oblique dorsal muscle may act synergically with the dorsal longitudinal muscle, but has a functionally antagonistic action in many orders. The dorsoventral muscles, running from tergum to sternum, act in opposition to the dorsal longitudinal muscles and supply power for the upstroke by lowering the anterior notal process relative to the pleural process. The oblique intersegmental muscle has a variety of roles but may have become an indirect wing depressor in the metathorax of Odonata (Clark, 1940).

The direct muscles can be divided into the basalar and subalar muscles, which act on the wing by virtue of the ligamentous attachment of these sclerites to the wing base, and those

4

muscles which are inserted directly on to the axillary sclerites. The mode of operation of the direct muscles varies greatly in different insects and has been worked out in only a few examples. The flexor muscle (fig. 3), however, has the same action in all wing-flexing insects, and this is conveniently described now. It runs from the pleuron to the middle of the 3rd axillary, often reaching this sclerite by a slender apodeme, and may be composed of several branches arising on the episternum, the pleural ridge and the epimeron. Its contraction rotates the 3rd axillary dorsally and inwards about the axis

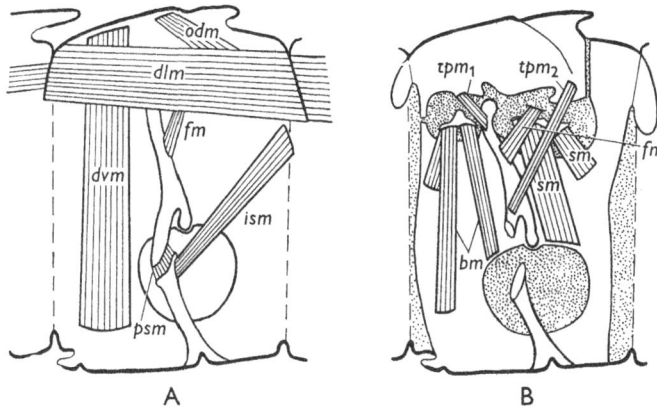

A   B

Fig. 3. Diagrammatic view of the muscles on the right side of a wing-bearing segment, seen from within. A, indirect and some lateral muscles; B, lateral muscles. *bm*, basalar muscles; *dlm*, dorsal longitudinal muscle; *fm*, flexor muscle; *ism*, oblique intersegmental muscle; *odm*, oblique dorsal muscle; *psm*, pleurosternal muscle; *sm*, subalar muscles; *tpm*, tergopleural muscles; *dvm*, dorsoventral muscle. (Redrawn from Snodgrass, 1935.)

formed by its articulations with the 2nd axillary and the posterior notal process (or the 4th axillary when present), thus carrying the vannal area of the wing dorsally and posteriorly by rotation about the 2nd axillary sclerite (fig. 2); folding of the wing along definite lines and some movement of the other axillary sclerites relative to each other and to the notal processes accompanies this complicated action.

In addition to these direct and indirect muscles whose general mode of action in the flight machine has been known for a long time, there are certain muscles, which may be called accessory indirect muscles, whose important role has only recently been demonstrated. The pleurosternal group is most commonly represented by a short but powerful muscle linking the pleural and sternal apophyses (fig. 3A). These structures form a system of internal bracing for the walls of the pterothorax, and in some insects the two apophyses are rigidly fused; where the muscle exists across the gap, its contraction influences the elasticity of the thoracic box, with corresponding changes in the insect's flight (see Chapter 2). Also in this category come a variety of tergopleural muscles, one of which, running from the posterior part of the scutum or scutellum to the pleural ridge, may be of considerable size in some orders. All these muscles influence flight by altering the relative position of moving parts or by changing the elastic properties of the pterothoracic box, rather than by direct action on the wings themselves.

Finally, it must be mentioned that leg muscles may become incorporated into the flight machinery. The coxa and even the trochanter of the leg are moved partly by muscles which have their origin on the tergum. When, as in the Diptera and some Coleoptera, parts of the coxae become rigidly attached to the wall of the thorax, such muscles may function as accessory dorsoventral muscles in the upstroke of the wing. Even when the coxa remains movable, the upper insertion of some of its muscles may be transferred to the wing or epipleural sclerites (basalare or subalare) and the muscles then have a double role; in the metathorax of a cicada the same muscle raises the hind legs into a position close to the body and also brings the hind wing forward into the flight position. Examples of other functions of muscles which are secondarily drawn into the flight machinery will be mentioned in later chapters.

CHAPTER 2

The Form and Mechanism of the Wing Beat

LIFT and propulsion are produced in the majority of insects by active movement of the wings and not solely by the air flow resulting from the insect's forward motion. Forward motion of the body as a whole may be necessary for flight, but in the most advanced fliers, which can hover or fly backwards and sideways, the analogy of a helicopter is closer than that of a conventional aeroplane. Unlike a helicopter the movement is oscillatory rather than rotary and the axis of rotation is horizontal rather than vertical, but in both types of flying machine changes in the angle of attack of the wing in different parts of the stroke are necessary to produce lift, propulsion and control.

KINEMATICS. The movement of the wings during flapping flight is an extremely complicated action involving, as well as elevation and depression, promotion and remotion (fore and aft movement), pronation and supination (twisting) and changes of shape by folding and buckling. The kinematics have been fully described in only one example, the locust *Schistocerca gregaria* (Jensen, 1956). Earlier studies using high-speed photography (Magnan, 1934) or reflecting markers glued to the wing (Marey, 1868*a*, *b*; Hollick, 1940) have provided only incomplete information which is inadequate for an aerodynamic analysis or for a full understanding of the mechanism of the articulation and the role of the various muscles. There is also the difficulty that the form of the wing beat is greatly affected by the 'relative wind' which includes a component due to the translational velocity of the insect as a whole; studies with fixed insects are therefore likely to be of limited value unless a wind tunnel is available and care is taken to ensure that the air flow is correctly adjusted to the conditions found in free flight.

The movement of the fore wing of *Schistocerca* during steady forward flight is shown in fig. 4. Relative to the insect the wing tip moves in an irregular loop; relative to the air it is an

7

irregular saw-tooth curve. The hind-wing movement is similar but not identical. At the mid-point of the wing the oscillatory component of the movement is smaller relative to the translational component, while at the base only the translation is significant. Fig. 4 also shows the twist and changes of wing section during the stroke in a particular set of experiments. At the base the inclination of the wing was constant and equal to the body

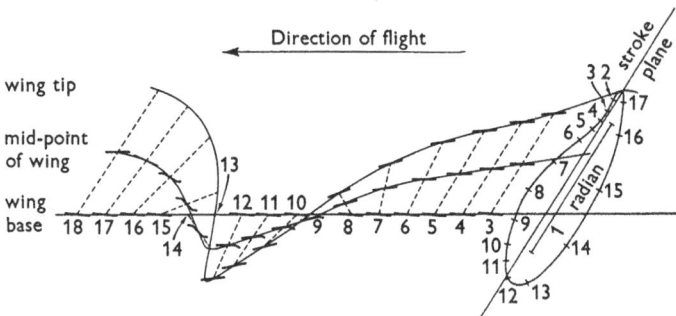

Fig. 4. The movement of the fore wing of *Schistocerca gregaria* (redrawn from figs. III, 5, III, 6 and III, 8 of Jensen, 1956). The closed curve shows the path of the wing tip relative to the insect. The open curves show the path relative to the air, on a projection designed to show the movement of the axis of the wing. Angles of attack are shown by the short lines, which also illustrate the changes in wing section at the mid-point of the wing.

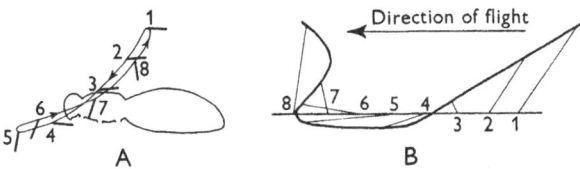

Fig. 5. A. Diagrammatic lateral view of the wing trajectory of *Volucella*, showing the changes in wing twisting during the stroke. B. The path of the wing tip through the air during forward flight. (Redrawn from Magnan, 1934.)

angle; from points 9 to 12 the middle of the wing was bent along the vannal fold to produce a flap; on the upstroke it was doubly bent into a Z-section. Comparable but less exact data for the wing-tip movements of *Volucella* (Diptera) in free flight are given by Magnan (1934) from analysis of high-speed photographs (fig. 5).

Studies using gold leaf and other reflecting materials on the wing tips have given information about the amplitude and form of the stroke in a wider range of insect types. In this way

Stellwaag (1916, fig. 51) was able to demonstrate changes in inclination of the major axis of wing movement (stroke plane) of *Apis mellifera* during flight manoeuvres; Vanderplank (1950) showed the same thing in *Glossina palpalis* (Diptera) by flash photography. Hollick (1940) demonstrated that there are changes both in the stroke amplitude and in the wing-tip curve when *Muscina stabulans* is placed in still air or in an air stream approximating in velocity to its normal forward movement (fig. 6). In general it may be said that the wing moves forwards and downwards with a positive angle of attack and then backwards and upwards with a large supination twist. Viewed from

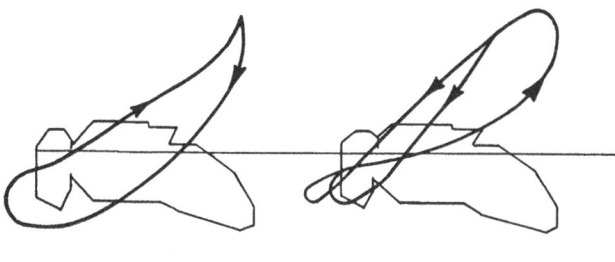

Still air   140 cm./sec.

Fig. 6. The path of the left wing tip of *Muscina stabulans* (Diptera) in still air and in an air stream (two curves showing the range of variation). (From Hollick, 1940.)

a direction at right angles to the stroke plane, the amplitude of beat varies in different types from 70° (*Aeshna*, Odonata) to 160° (*Lucanus* and other beetles) (Magnan, 1934).

Of all the parameters involved in a description of insect flight the most readily measured is the frequency of beat. Chadwick (1953) devotes considerable space to this aspect of the problem and discusses the physiological and environmental factors which affect it. Sotavalta (1947) gives a comprehensive table of data, obtained by aural estimation of the flight tone. The frequency of wing beat varies from 5 per sec. in *Papilio machaon* (Lepidoptera) to about 1000 per sec. in *Forcipomyia* (Diptera).

Within a single wing stroke it has often been assumed that the motion is sinusoidal (Sotavalta, 1952). In fact, the observed wing-tip curves (fig. 6) and even more the measurements which have been made from high-speed cinematographs (fig. 4) show that the velocity of wing movement varies through the cycle in a manner far from sinusoidal and incapable of description by

9

any simple mathematical formula. It has even been claimed (Vanderplank, 1950) that in *Glossina* there is a definite pause at the top and bottom of the strokes, so that each up- and down-stroke occupies only one-eighth of the total duration of the cycle. Here, as with the nature of the angular motion, it is clear that there is no short cut to an accurate kinematic treatment of each insect type on which it is intended to conduct quantitative aerodynamic or physiological investigations.

The role of the various articular sclerites in the production of this complicated pattern of movement in time and space cannot be described in terms which are more than approximately true for insects in general, and even in particular cases there are few published accounts which are at all complete. A comprehensive review of functional morphology will not be attempted here, but

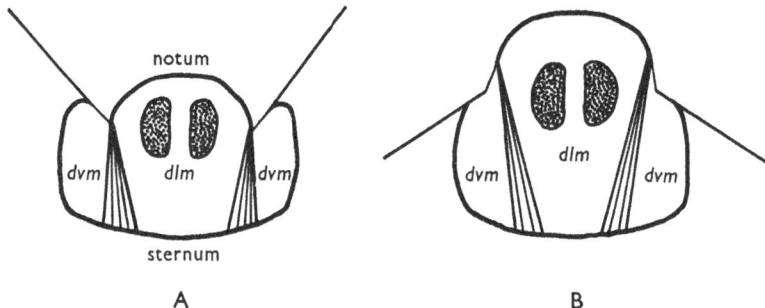

Fig. 7. The classical diagram showing, in transverse section of the thorax, the mechanism of the indirect flight muscles according to the scheme of Chabrier (1822). A. Top of upstroke. B. Bottom of downstroke. *dlm*, dorsal longitudinal muscles; *dvm*, dorsoventral muscles. (Redrawn from Magnan, 1934.)

certain features must be discussed because they are important for an understanding of the physiology of flight. These relate, in the first instance, to the mechanism whereby the forces generated by contraction of the indirect muscles are transmitted to the wings.

ACTION OF THE INDIRECT MUSCLES

DIPTERA. Textbooks of entomology usually give a description of the mode of action of the indirect muscles which is essentially that of Chabrier (1822). This simple picture of an up-and-down movement of the notum relative to the sternum and pleuron (fig. 7) is correct only for certain orders, and even there ignores the inherent elasticity of the thoracic skeleton. Weis-Fogh (unpublished) has measured the elastic deformations in the thorax of *Schistocerca* and has evaluated the considerable contribution which elastic forces make in the generation of the wing strokes. His conclusions are generally applicable to the majority of insects and have been illustrated by models. In the simplified model of fig. 8A the wing is assumed to be a rigid rod articulating with the notum and pleuron, the thorax having a natural elasticity (represented by the coil spring) which produces an equilibrium position with the wings depressed. A force applied to the top of the notum raises the wings, as in the

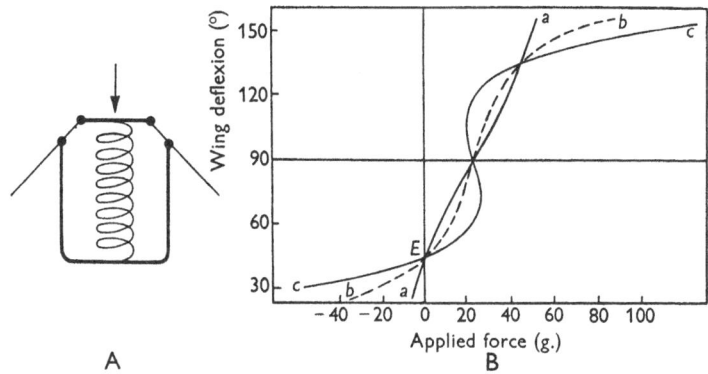

A   B

Fig. 8. A. Simplified model of the insect thorax, demonstrating the effect of elastic forces on the wing movement. B. Relationship in the model between vertically applied force and wing deflexion; *a*, with no lateral stiffness; *b*, *c*, with increased lateral stiffness; *E*, equilibrium position of model shown in A. (Redrawn from Weis-Fogh, unpublished.)

Chabrier picture (fig. 7), but lateral elasticity is introduced by the bending of the frame representing the sternum and pleura, and the relationship between angular wing deflexion and applied force depends on the relative magnitude of this elastic force. If the lateral stiffness is slight, the curve approximates to a straight line (*a*, fig. 8B); as lateral stiffness is increased there comes a point when the curve becomes re-entrant (*c*, fig. 8B) and the system is then bistable; the wing 'clicks' suddenly from an up to a down position when the applied force reaches a critical value.

    Lateral stiffness of the pleural walls of the thoracic cavity is achieved in nearly all insects by the reinforcement of the pleural ridge (*PS*, fig. 1) and by the internal pleural and sternal apophyses which are often linked by the powerful pleurosternal muscles (*psm*, fig. 3 A). Contraction of the pleurosternal muscles provides an adjustment of the lateral stiffness and therefore of the form of coupling between the indirect muscles and the wings. In certain cases the apophyses are fused (mesothorax of Coleoptera and Sphingidae; metathorax of Hymenoptera), producing a strong but constant lateral stiffness. Where the force for the downstroke is provided mainly by direct muscles, these may be so arranged that their pull increases the lateral stiffness as well as moving the wings (Odonata, p. 21; metathorax of Coleoptera, p. 30). The widespread occurrence of features of this nature emphasizes the importance of lateral stiffness in the mechanism of wing movement.

The model of fig. 8 A is too simplified to be an accurate representation of any actual insect, although it illustrates well the essential mechanics of the thoracic system. There is usually some movement between the axillary sclerites at the wing base, and in its extreme manifestation this movement can allow the wings to beat without much vertical displacement of the notum relative to the sternum. Such a system was described by Boettiger and Furshpan (1950, 1951, 1952) for the blow-fly, *Sarcophaga bullata*, in the first account of a 'click mechanism'. These authors found that, although in a fly anaesthetized with carbon dioxide the wings can easily be pushed into any position, flies anaesthetized with carbon tetrachloride (CCl_4) assume a condition in which the wings can rest only at the extremes of their range of movement, giving a bistable state similar to that of the model with large lateral stiffness. Boettiger and Furshpan demonstrated that the pleurosternal muscle contracts on treatment with CCl_4, and were able to show that the click action of the articulation is a normal feature of the wing beat and persists even in wingless flies executing flight movements. Their description of the mechanism must be given in full, since it is valid for all the higher Diptera which have been used for so many flight studies; variations containing some of the same features are also found in other orders.

In the thorax of a fly the tergal, pleural and sternal sclerites are fused to give a strong, elastic structure apart from certain well-defined regions where the cuticle remains thin and flexible. The powerful dorsal longitudinal indirect muscles run right across the thoracic box, and their pull at the posterior end is transmitted from the postphragma through the postnotum to a point (i, fig. 9A) at which the postnotum articulates with the scutellum by a hinge joint. Forward and laterally from this point on each side an arm (called by Boettiger and Furshpan the 'scutellar lever') extends to the region of the wing base and carries on its ventral end the posterior notal process; to this process, as usual, articulates the 3rd axillary sclerite, but this is not concerned in the click mechanism and is not shown in fig. 9. On the extreme anterior end of the scutellar lever a process articulates also with the 1st axillary sclerite at point z (fig. 9B, C), and it is by forces transmitted through this point that the wing strokes are initiated.

The scutellum is joined to the scutum along the scutoscutellar suture which allows some bending; the rest of the scutellum, with its lever, is separated from scutum and postnotum by thin, flexible cuticle apart from the articulation at point i. There are two grooves kf and on which allow some relative movement of the lateral parts of the thorax, and the anterior notal process (ANP, y, fig. 9) is borne on a lateral shelf, the anterior parascutum (p, fig. 9), which can hinge on the scutum along the line fg (fig. 9A).

Contraction of the dorsal longitudinal muscles, producing a forward push on the scutellum at point i, raises the anterior end of the scutellar lever and depresses the posterior end of the scutellum; contraction of the dorsoventral muscles pivots the whole scutum about the line jk (fig. 9A) and induces the opposite movement. The two sets of muscles are not, however, entirely antagonistic in their action, since both tend to produce a lateral expansion of the notum at point f.

The click mechanism can now be understood. If the wing is in the UP position (fig. 9B) the first effect of the contraction of the dorsal longitudinal muscle is to force apart the two points f on each side. This lateral force is transmitted through the parascutal shelf (p) and the 1st and 2nd axillary sclerites to the pleural wing process, but here it is resisted if the pleurosternal muscles are also contracted; the only wing movement which can occur is a very slight one due to the forcing further out of line of the three articular points (x, y, f), at the junction respectively of the pleuron with the 2nd axillary, of the 1st axillary with the parascutum (the anterior notal process) and of the parascutum

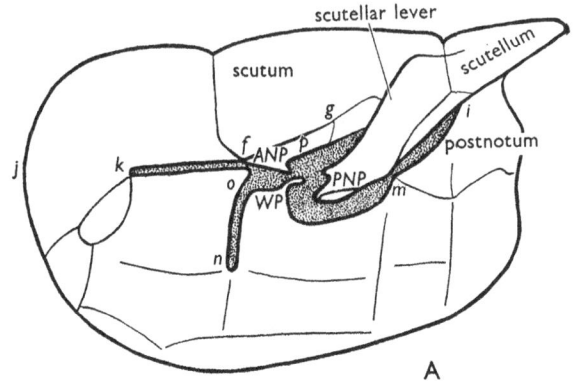

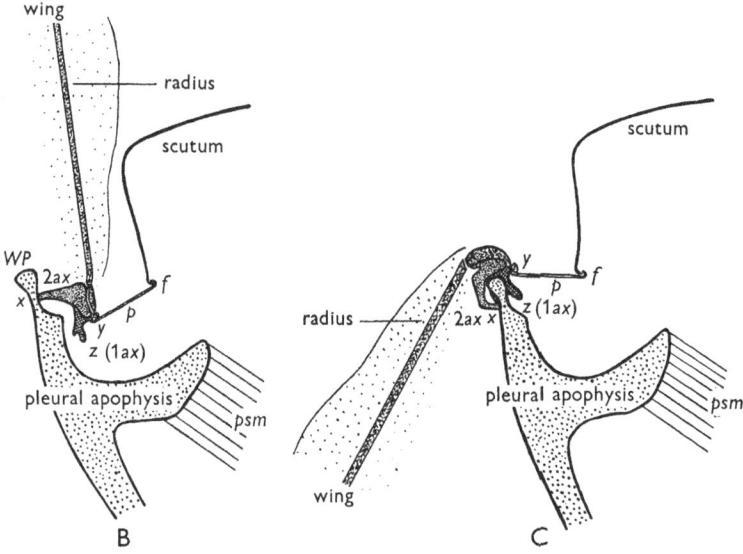

Fig. 9. The click mechanism of the fly wing (redrawn from Boettiger and Furshpan, 1952). A. Diagrammatic lateral view of the thorax of *Sarcophaga bullata*, from the left side. B, C. Diagrammatic transverse sections of the left wing-base of *Sarcophaga*, showing the position of the articular pieces at the start of the downstroke (*B*) and upstroke (*C*). *ANP*, anterior notal process; *PNP*, posterior notal process; *WP*, pleural wing process; *psm*, pleurosternal muscle; *1 ax*, *2 ax*, 1st and 2nd axillary sclerites; other labels are explained in the text.

with the scutum. At the same time the movement of the scutellar lever is applying an upward force to point z, where its process articulates with the 1st axillary. When this force becomes sufficient to carry point y nearly to the line joining x and f, the elastic energy stored in the notum is released by a rapid rotation of the 2nd axillary about the pleural wing process and the wing 'clicks' into its DOWN position (fig. 9C); the 1st and 2nd axillary sclerites (whose relative movement is in a different plane from that of the paper in fig. 9B,C) together with the parascutal shelf form a toggle joint which assists the click action. On the upstroke, again, the first effect of contraction of the indirect (dorsoventral) muscles is to force apart the two points f; and, again, little movement can at first take place (fig. 9C), since point y is above the line xf; when the downward force applied by the scutellar lever to point z again reaches the critical value the wing clicks back into the UP position. It is clear that this mechanism only works if the lateral movement of point f is resisted by contraction of the pleurosternal muscles. This accounts for the failure of earlier workers to detect the click action, since only with certain anaesthetics (e.g. CCl_4) are this and the other accessory muscles put into tonic contraction.

Boettiger and Furshpan (1952), by recording the movement of the scutellum, were able to show that at the start of flight the first event is a slow downward movement of the wings to the position shown in fig. 9C. The initial fast upstroke is probably produced by contraction of the tergotrochanteral muscle (part of the dorsoventral indirect muscle complex: see p. 38); this movement then starts the rhythmic mechanism of the wing beat.

The arrangement of the basal articulations in a fly ensures not only that the force of contraction of the indirect muscles is transmitted to the wing, but also that movements of pronation and supination (twisting) occur at the correct phase of the cycle to produce lift and propulsion. This is achieved by relative movement between the 1st and 2nd axillary sclerites and by the fact that point z (fig. 9B,C), at which the scutellar lever exerts its vertical force, lies posterior to point y, the mid-point of the toggle. The explanation of the automatic twisting of the wing was apparent to Boettiger and Furshpan (1952), but their account is difficult to follow without the aid of a working model (fig. 10).

The distortion of the thoracic box by the contraction of the dorsal longitudinal and dorsoventral indirect muscles is

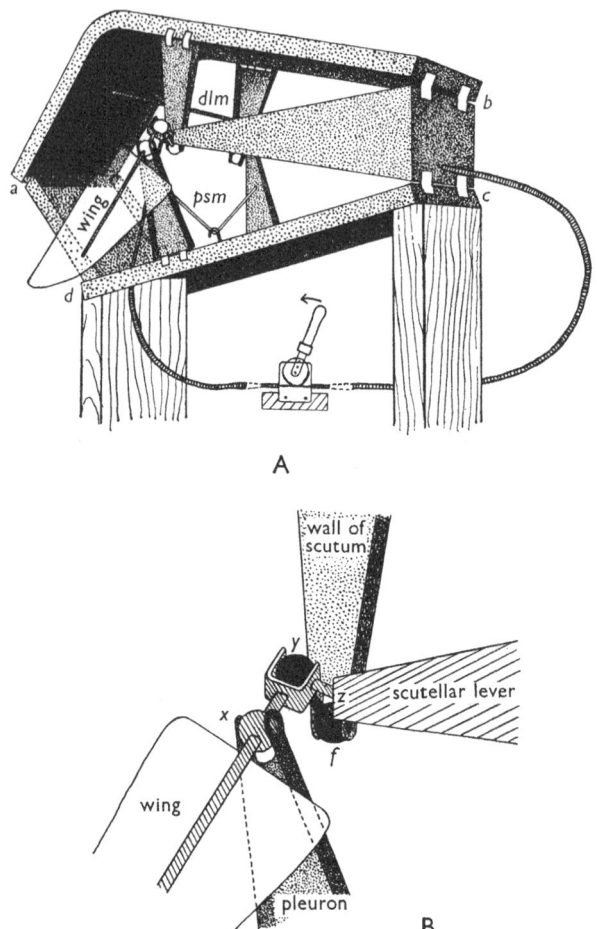

Fig. 10. A. Working model to illustrate the click mechanism of the fly wing. *dlm*, wire representing the dorsal longitudinal muscle; *psm*, elastic representing the pleurosternal muscle. B. Enlarged view of the articular mechanism. Labels are explained in the text.

represented in the model by the lozenge distortion of the trapezium *abcd*, the four plates of which it is constructed being hinged by spring-steel strips. The plate *bc*, representing the postphragma, postnotum and scutellum, carries on each side a lateral arm, the scutellar lever, whose tip is provided with a short vertical slot (not shown in the figure). The pleuron and the lateral wall of the scutum are represented by flanged plates which are also connected by steel strips to allow some movement. The top of the pleural plate and the bottom of the notal plate carry simple universal joints made of a spherical polythene bead held between two circular holes in the flanges of the plates; these represent respectively the pleural wing process (x) and the basal hinge of the parascutum (f). Details of the articulation are shown in fig. 10B. The single wing vein is carried through the universal joint of the pleural wing process to another universal joint corresponding to the anterior notal process (y), and also rigid with this assembly is a backwardly directed peg representing the backwardly directed arm of the 1st axillary sclerite; this fits into the slot in the end of the scutellar lever to correspond to point z (fig. 9B). From the second universal joint (point y) a short rod bearing two polythene beads links up with the lateral wall of the scutum and represents the anterior parascutum (p, fig. 9B). The essential moving parts are thus copied in the model apart from the fact that a single rigid assembly with two universal joints replaces the 1st and 2nd axillary sclerites; the combination of up-and-down movement of the wing with twisting is achieved in the fly with some relative movement of these two pieces, and there is probably some control of the stroke by direct muscles which has not been imitated in the model.

The model is worked by Bowden wire cables placed in the position of the indirect muscles, and it is shown in fig. 10A at the beginning of the upstroke. Shortening of the wire representing the dorsoventral muscles tends to depress the tip of the scutellar lever (point z), and, since the pressure on the toggle (xyf) initially prevents any upstroke movement, the whole wing twists into a supinated position. As the applied force increases the wing suddenly snaps into the UP position. Pronation is similarly and automatically produced by the upward pressure of the tip of the scutellar lever before the movement of the downstroke; apart from the rigidity of the whole wing the twisting during the stroke cycle is well demonstrated. Re-examination of fig. 9B, C will show that a similar action should occur in the fly.

The action of the pleurosternal muscles is easily imitated. They are represented in the model by a band of elastic, attached to the two pleural plates and looped round a hook representing the sternal apophysis. With the elastic off the hook (muscles relaxed) there is little 'click' action in the wing movement, but if the elastic is stretched (muscles contracted; for example, by

CCl$_4$) the click becomes powerful enough to be felt as a sudden removal of resistance to the movement of the operating lever.

Boettiger and Furshpan (1952) make two further points about the operation of the wing cycle in the CCl$_4$-anaesthetized fly. During the initial phase of contraction of the indirect muscles when the cuticle and the articulation are being strained without movement of the toggle joint, there is some relative movement of the anterior notal and pleural wing processes in a mainly horizontal direction, which produces a fore and aft movement of the wing tip. In normal flight this is out of phase with the stroke and is responsible for the irregular elliptical or figure-of-eight-shaped path of the wing tip; the strict parallelogram construction of the model reduces this effect, but it is present nevertheless in the same sense as in the fly. Secondly, they rightly point out that limitation of the amplitude of the stroke is necessary at the end of the 'clicks'. The upstroke is limited in the fly by the fact that the scutellar lever comes into contact with the pleural sclerites at point m (fig. 9A); this is an invariant stop, incapable of adjustment. The downstroke is limited by the nature of the articulation between the 1st and 2nd axillary sclerites, which may be varied by means of direct muscles. It is significant that Hollick (1940) found that control of stroke amplitude in *Muscina* is achieved by variation of the lower limit of the stroke (fig. 46). These stops are not correctly imitated in the model.

Direct proof of the presence of a click action in the wing articulation has been obtained for the higher Diptera, for the metathorax of *Schistocerca*, and for the metathorax of a beetle (p. 30), and there are indications from morphology that it may be present in many other types; it would be unwise, therefore, to assume that Chabrier's simple picture (fig. 7) is a complete representation of the action of the indirect muscles in any insect.

HYMENOPTERA. There has been no recent functional investigation of the mechanics of wing motion in any hymenopteran, but our knowledge of the morphology of the thorax of the hive-bee, *Apis mellifera*, is perhaps more complete than that of any other insect, through the work of Snodgrass (1910, 1925, 1942, 1956). This author's interpretation of the role of the various muscles is, however, incorrect; a re-examination of the hive-bee and the large carpenter bee *Xylocopa tenuiscapa* suggests that Stellwaag (1910) has given a more accurate account of the mode of opera-

18

tion of the large mesothoracic indirect muscles, which alone are capable of the high speed of contraction required for the production of the wing beat (Chapter 3). As in the Diptera, these muscles are responsible for the basic movements of pronation and supination as well as for depression and elevation of the wing, and in the Hymenoptera they achieve this by producing relative movement between the anterior and posterior notal processes. This is the method of wing movement first noticed by Janet (1899) and contrasted by him with the anterior notal

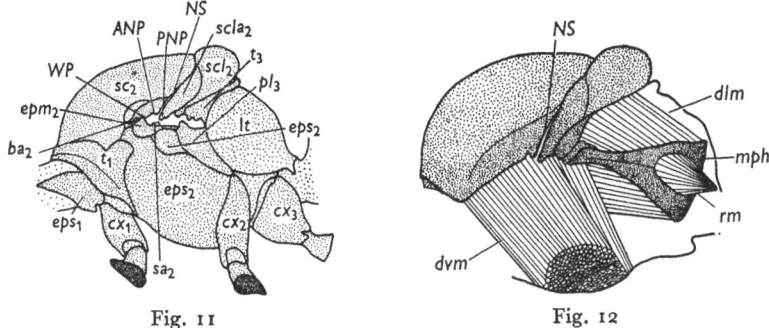

Fig. 11   Fig. 12

Fig. 11. Semi-diagrammatic lateral view of the left side of the thorax of *Apis mellifera* (worker). *ANP*, anterior notal process; *NS*, notal suture; *PNP*, posterior notal process; *WP*, pleural wing process; *ba*, basalare of mesothorax; cx_1, cx_2, cx_3, coxae of legs; epm_2, mesepimeron; eps_1, proepisternum; eps_2, two parts of mesepisternum; pl_3, metapleuron; sa_2, subalare of mesothorax; sc_2, mesoscutum; scl_2, mesoscutellum; $scla_2$, mesoscutellar arm; t_1, protergum; t_3, metatergum; *It*, tergum of 1st abdominal segment (propodeum). (Redrawn from Snodgrass, 1925.)

Fig. 12. Semi-diagrammatic lateral view of the thorax of *Apis mellifera* (drone), with the indirect flight muscles exposed. *NS*, notal suture; *dlm*, dorsal longitudinal muscle; *mph*, mesophragma; *rm*, retractor muscle of mesophragma (dorsal longitudinal muscle of metathorax); *dvm*, dorsoventral muscles. (Redrawn from Snodgrass, 1942.)

(pleural) movement described by Chabrier (1822). Snodgrass (1910) gives an anatomical account of the evolution in the Hymenoptera of the transverse mesonotal suture which makes possible the relative movement of anterior and posterior notal processes (*NS*, figs. 11, 12).

The sclerotization of the bee thorax is illustrated in fig. 11 and the arrangement of the indirect muscles in fig. 12. The mesophragma is an internal structure lying far back in the propodeum and transmitting the force of contraction of the dorsal longitudinal muscle to the base of the scutellar arm which carries the posterior notal process. The scutellum can turn about a horizontal elastic hinge across the notum, and the presence of the mesonotal suture allows the arm to slide forward and upward, under (median to) the lateral wall of the scutum bearing the anterior notal process, until the movement is limited by a skeletal stop. This motion is transmitted to the wing by the 1st axillary, which is shown flat in fig. 13 and in its natural

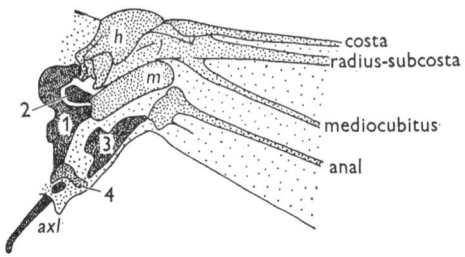

Fig. 13. Semi-diagrammatic view of the flattened base of the right fore wing of *Apis mellifera*; *axl*, axillary lever; *h*, humeral complex; *m*, median plate; 1, 2, 3, 4, axillary sclerites. (Redrawn from Snodgrass, 1942.)

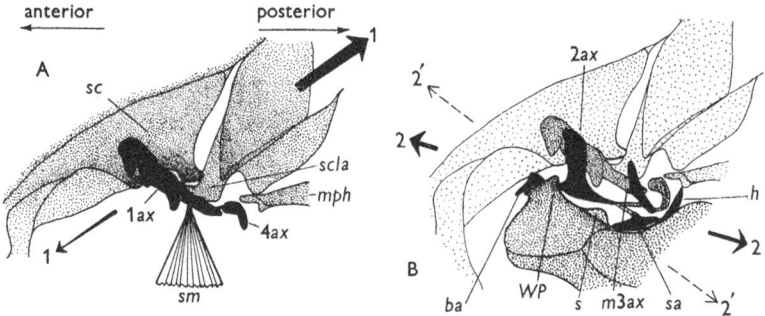

Fig. 14. A, B, external views of the natural positions of certain components of the basal articulation of the left fore wing of *Apis mellifera*; *ba*, basalare; *h*, rod-like pleural sclerite; *mph*, mesophragma; *m3 ax*, point of attachment of folding muscles to 3rd axillary sclerite; *s*, socket on pleuron; *sa*, subalare; *sc*, scutum; *scla*, scutellar arm; *sm*, scutellar muscle; 1 *ax*, 2 *ax*, 4 *ax*, axillary sclerites; 1–1, axis of turning of 1st axillary; 2–2, 2′–2′, axes of turning of 2nd axillary sclerite on down- and upstrokes. (Original.)

position in fig. 14. This sclerite rests in a smooth cup in the lateral wall of the scutum (anterior notal process) and has a long posterior arm which is attached by ligaments to the end of the scutellar arm and to the 4th axillary; when the mesonotal suture closes it turns about the axis 1–1 shown in fig. 14A, which is inclined at about 45° to the plane of the drawing (upper end of axis towards the reader). The upper end of the 1st axillary (which projects out of the plane of the drawing) turns forwards and downwards; since this is attached to the humeral complex (fig. 13), the wing is thereby pronated as well as depressed. The upper part of the 2nd axillary (fig. 14B) rests in a cup of the 1st axillary, and the movement of the scutellar arm rotates the 1st axillary round the 2nd axillary which pivots forward about the pleural wing process, lifting its posterior arm into the position shown in fig. 14B. When further contraction of the dorsal longitudinal muscle moves the anterior notal process outwards and upwards the 2nd axillary turns about the axis 2–2 (fig. 14B) and the wing is depressed in a pronated attitude.

At the beginning of the upstroke, contraction of the vertical indirect muscles moves the scutum downwards and backwards, which has the effect of reversing the movements of the 1st and 2nd axillaries. The backward movement of the notum, transmitted through the 1st axillary, pivots the 2nd axillary backwards about the pleural wing process until its posterior arm engages with a cup-shaped socket (s, fig. 14B) on the pleuron, thereby changing its axis of rotation to that of the interrupted arrow ($2'–2'$, fig. 14B) and supinating the wing before the upstroke. The arrangement of the indirect muscles and sclerites thus ensures that pronation and supination occur in the correct phasing to the up- and downstrokes.

RELATIVE MOTION OF FORE AND HIND WINGS

ODONATA. The most primitive insects have the pterothoracic segments of approximately equal size; homologous flight muscles, both direct and indirect, can be found in the Orthoptera and other orders with equally developed wings. The attachment of both dorsal longitudinal muscles to the mesophragma necessarily produces some mechanical interaction between the meso- and metathoracic segments, and there is present in the Orthoptera a considerable phase difference between the motion of the two pairs of wings (Weis-Fogh, 1956a). In those higher orders which rely on the indirect muscles to provide the main motive power for flight there is a tendency, in several independent evolutionary lines, for one of the two pterothoracic segments to be reduced; this is most marked in the Homoptera Auchenorrhyncha, Hymenoptera and Diptera in which the mesothoracic indirect muscles motivate both pairs of wings, and in the Coleoptera, where it is the metathoracic muscles which are fully developed. A reduction in size of the passively moved hind wings occurs in the former orders, with hook coupling to the front pair, while in the Coleoptera the aerodynamic function of the elytra is limited to the provision of fixed lift-generating surfaces (p. 97).

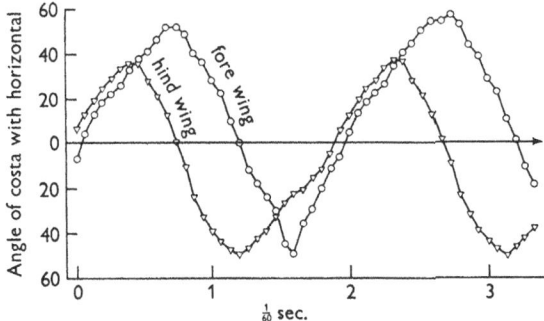

Fig. 15. The angular motion of the wings of a dragonfly, *Ladona exusta*, analysed by high-speed cinematography of a tethered insect. Note that the *x*-axis is a scale of time, not distance. (Redrawn from Chadwick, 1940.)

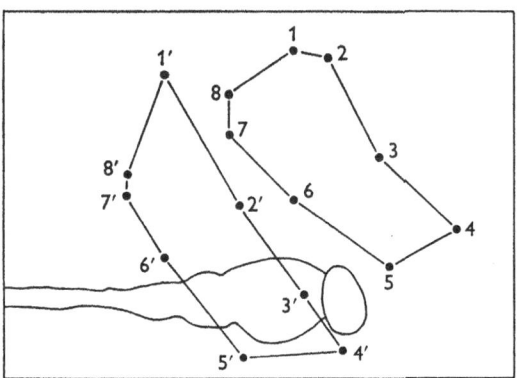

Fig. 16. Projection on to the vertical fore-and-aft plane of the wing-tip motion of a dragonfly; the numbers refer to instants of time during the cycle. (Redrawn from Magnan, 1934, who made the measurements on a model set in positions similar to those shown in high-speed cinematographs.)

In the Odonata, with equally developed fore and hind wings, a similar phase difference is found to that of the locust, the hind wing reaching the bottom of its stroke while the fore wing is still moving downwards with maximum velocity (figs. 15, 16). The thorax of a dragonfly differs considerably from that of other insects in the relative size of its constituent parts, but the

22

homology of the various sclerites and muscles can be clearly traced (Clark, 1940; Chao, 1953). The meso- and metathoracic terga consist of small, loosely articulated plates, coupled to the wing base through the 1st axillary sclerite and large humeral and axillary plates; the other axillary sclerites are not present as separate pieces. The pleural ridges are elongated and internally strengthened (Sargent, 1937), forming stout pillars on each side parallel to the main wing muscles and providing a curved compression strut to balance their pull (fig. 17 B); the wing turns about the pleural wing process as a simple lever and its amplitude of movement is small. Dorsal longitudinal indirect muscles

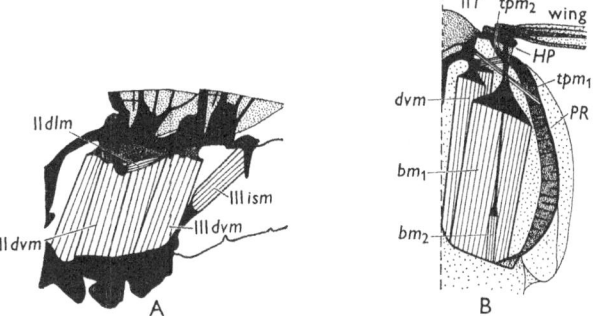

Fig. 17. A, internal view of the right side of the pterothorax of *Anax junius*. II *dlm*, mesothoracic dorsal longitudinal muscle; II *dvm*, III *dvm*, meso- and metathoracic dorsoventral muscles; III *ism*, metathoracic oblique intersegmental muscle. (Redrawn from Clark, 1940.) B, anterior view of the left mesothorax of *Anax imperator*. *HP*, humeral plate; *PR*, pleural ridge; II *T*, mesotergum; bm_1, bm_2, first and second basalar muscles; *dvm*, dorsoventral muscle; tpm_1, tpm_2, tergopleural muscles. (Original.)

are reduced or absent (II *dlm*, fig. 17 A). The upstroke is produced by the large dorsoventral indirect muscles (*dvm*, fig. 17 A, B), and the downstroke by the large 1st basalar and subalar muscles; in the metathorax, Clark (1940) states that the indirect oblique intersegmental muscle (III *ism*, fig. 17 A) is a depressor, but this requires confirmation. Two small tergopleural muscles and the small 2nd basalar and subalar muscles (fig. 17 B) appear to control the form of the wing beat, and the former may increase the lateral stiffness of the pterothorax. The pleural and sternal apophyses are fused to provide a transverse bridge at the base of the pleural struts, and the arrangement of the large wing muscles in relation to the curved struts is such that any simultaneous tension in both depressor and levator sets will produce an inward pressure from the pleural wing process, perhaps generating some click action in the wing movement (fig. 17B).

According to Clark (1940) 'there are but two cardinal movements in Odonata, elevation and depression; in addition the wing rotates or pronates on its long axis'. Magnan's (1934) plots of the path of the wing tip (fig. 16) do not support this view, since there is also considerable fore-and-aft movement out of phase with the elevation and depression. In a system in which the downstroke is produced by the pull of large direct muscles inserted on the anterior humeral and the posterior axillary plates, these complicated wing-tip paths might be

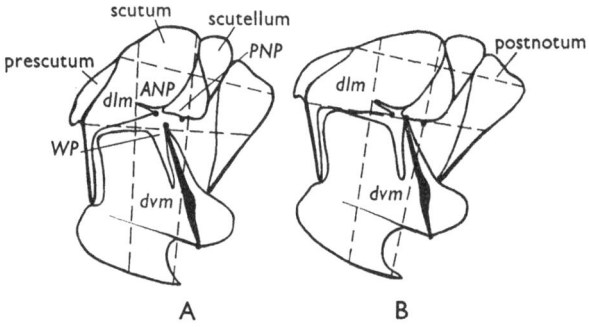

Fig. 18. Diagrammatic lateral view of the left side of the mesothorax of *Aphis fabae*. A, wing DOWN; B, wing UP. *ANP*, anterior notal process; *PNP*, posterior notal process; *WP*, pleural wing process; *dlm*, dorsal longitudinal muscle; *dvm*, dorso-ventral muscle. (Redrawn from Weber, 1930.)

achieved by differential timing of the contraction of the four muscle sets, but the mutual interaction of notal movements could also produce some antero-posterior displacement. More accurate functional studies are required on the Odonata before it will be possible to advance our knowledge of the wing mechanism of this order.

HEMIPTERA. This order displays a considerable diversity of form in the wings and wing articulations, and is of special interest in view of Tiegs's (1955) work on the flight muscles (p. 37). Some very detailed studies of certain types have been made by Weber (1928, 1929, 1930). Apart from the Aleurodidae, which have well-developed hind wings, there is a tendency for the main muscles to be concentrated in the mesothorax, with wing-coupling mechanisms to transmit the power for the beat; this reaches its limit in male Coccidae which have the hind wings filamentous or absent, but even a cicada can fly satisfactorily without the hind wings (Shen and Young, 1943). The wing articulation of an aphid (fig. 18) is an almost perfect example of the generalized type of indirect muscle system, distortion of the whole thoracic box producing a relative movement of the pleural and notal wing processes. A peculiarity in

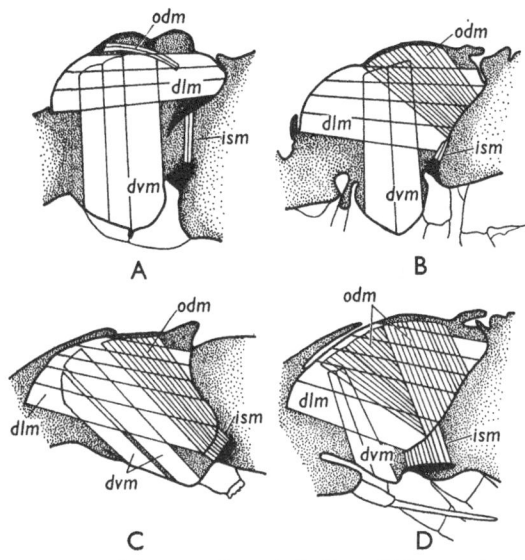

Fig. 19. Arrangement of the mesothoracic indirect flight muscles in Homoptera. A, *Aphis fabae*; B, *Psylla mali*. (Redrawn from Weber, 1930.) C, *Cyclochila australasiae* (Cicadidae); D, *Erythroneura ix* (Jassidae). (Redrawn from Tiegs, 1955.) *dlm*, dorsal longitudinal muscle; *dvm*, dorsoventral muscle; *ism*, oblique intersegmental muscle; *odm*, oblique dorsal muscle. Not to scale.

aphids is the presence of a modified basalare linking the preepisternum and the pleural wing process in such a way that contraction of the well-developed anterior pleural muscles brings the wing forward into the flight position; this would seem to be correlated with the ability of these insects to hold their wings extended without up-and-down movement (Kennedy, J. S., personal communication), a habit which may assist in their dispersal by air currents in the atmosphere.

In the arrangement of the indirect flight muscles, the Hemiptera can be divided into two groups (excluding the Aleurodidae). In the Aphidae, Coccidae and the Heteroptera, the dorsoventral indirect muscles are well developed, and the oblique dorsal muscle (*odm*, fig.19 A) is small; according to Weber (1930) its main function in the male coccid is the unfolding of the wings. In the Psyllidae and the Homoptera Auchenorrhyncha, the dorsoventral muscles are smaller and the oblique dorsal muscle is large (*odm*, fig 19 B, C). Owing to the great development of the postphragma the posterior insertion of the oblique dorsal muscle is carried down so far that its fibres come to lie almost parallel to those of the dorsoventral muscle; rigid coupling of the sides of the postphragma to the pleuron and the more lateral insertion of the oblique muscle on the notum make it into a wing levator. The arrangement is well seen in a cicada (fig. 19 C), where it was described by Snodgrass (1921). Accompanying the enlargement of the postphragma is a change of function also for the oblique intersegmental muscle (*ism*, fig. 19; compare *ism*, fig. 3 A), which becomes in cicadas merely a powerful strengthening muscle for the lower end of the phragma and almost reverses the normal alignment of its fibres.

Measurements have been made by a number of authors of wing-beat frequency in Hemiptera; the values range from 30 per sec. for *Erythroneura ix* (Jassidae) (Tiegs, 1955) through 40 per sec. for *Platypleura capitata* (Cicadidae) (Pringle, unpublished) and 85 per sec. for *Cryptomyzus galeopsidis* (Aphidae) to 100–140 per sec. for various Psyllidae and Heteroptera (Sotavalta, 1947). Nowhere in the order are frequencies found as high as those in the Diptera and Hymenoptera.

COLEOPTERA. The Coleoptera and the related Strepsiptera are the only orders of insects in which the metathorax and the hind wings have greater importance in flight than the mesothorax. The elytra, though they have an important role in the aerodynamics of flight, are not actively moved up and down, their small motion resulting from the thoracic vibration transmitted forward from the metathorax (Stellwaag, 1914). In the majority of beetles the elytra are held during flight in a position from 30° to 45° above the horizontal, where they are maintained by the tonic contraction of indirect muscles and by a system of self-locking apodemes at the base; the complicated movements of unfolding, involving both direct and indirect muscles, are fully described by Stellwaag. An exceptional arrangement is found in the Cetoniinae, which raise the elytra at the start of flight only sufficiently to allow the wings to be

26

drawn from beneath them, wing movement being made possible by a lateral emargination of the antero-lateral edge of the elytra.

The stroke angle of the wings in flight is larger in beetles than in most insects; Magnan (1934) gives a figure of 160° for *Lucanus cervus* and the full 180° is probably reached in some species. The wing-beat frequency may be high for the size of the insect, reaching 175 per sec. in *Attagenus schäfferi* (Dermestidae) (Sotavalta, 1947). No accurate observations have been made of the wing-tip path or other details of the kinematics, but it is clear that in many cases the wings approach more nearly to one another at the bottom of the stroke than is usually the case; the mean position may be below the horizontal.

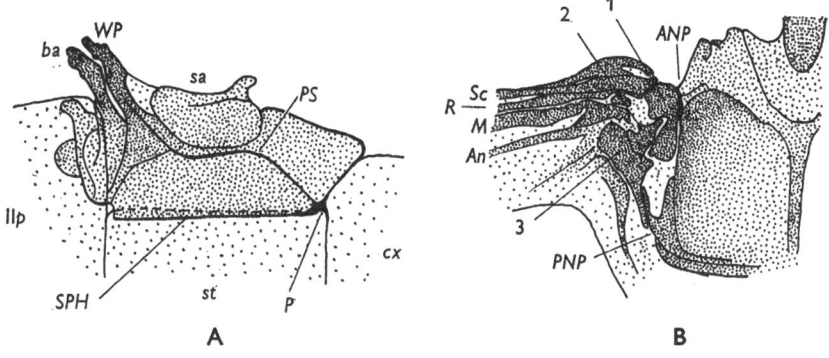

Fig. 20. The wing articulation of beetles. A, external view of left pleural sclerites of *Melolontha melolontha* (modified from Snodgrass, 1909). B, notum and left wing-base of *Acrocinus longimanus* (redrawn from Rüschkamp, 1927). *ANP*, anterior notal process; *ba*, wing process of basalar sclerite; *cx*, coxa of IIIrd leg; *P*, pivot point of sternopleural hinge; *PNP*, posterior notal process; *PS*, pleural suture; *sa*, subalar sclerite; *SPH*, sternopleural hinge; *st*, sternum; *WP*, pleural wing process; II*p*, mesopleuron; 1, 2, 3, axillary sclerites. Wing veins; *An*, anal; *M*, media; *R*, radius; *Sc*, subcosta.

Details of the pleural sclerites and of the wing articulation are given in fig. 20 and of the flight muscles in fig. 21. The pleural ridge runs horizontally back from the wing process to the (often rigid) coxa, and there is a remarkable development of the basalar region forming what appears to be almost a second wing process (*ba*, fig. 20), which locks between costa and sub-costa when the wing is folded and is linked with the apodemal cap of the basalar muscle. The sternum is heavily sclerotized and is joined to the pleuron by a long horizontal hinge (*SPH*, fig. 20A)

27

which allows some vertical movement about the pivot at its posterior end (*P*, fig. 20A). The metatergum bears large pre- and post-phragmata; the lateral regions of the scutum on each side are less reinforced than in many insects and are capable of some independent movement. The posterior notal process is carried on a wide lateral shelf (*PNP*, fig. 20B). The three axillary sclerites have the normal relationship to each other and to the notal and pleural wing processes, and the wing-folding mechanism shows no unusual features except that, distally on the wing, a transverse fold comes into action when the basal veins are compressed one upon another by the movement induced by contraction of the 3rd axillary muscles.

The indirect flight muscles (fig. 21A) consist, as usual, of the dorsal longitudinal between the metanotal phragmata, an oblique dorsal muscle ('musculus lateralis metathoracic tertius' of Stellwaag, 1914; 'lateralis posterior' of Rüschkamp, 1927), whose posterior insertion is carried down so far on the postphragma that it becomes a wing levator muscle, and tergosternal and tergocoxal dorsoventral muscles. Of the direct muscles (fig. 21B), the basalar and subalar are extremely well developed, as in Orthoptera, and are inserted dorsally on apodemal discs resembling those of Odonata. The subalar disc (regarded by Snodgrass (1909), as a portion of the epimeron) is joined by ligamentous cuticle to the posterior end of the 2nd axillary sclerite near the point where it articulates with the 3rd axillary; the basalar sclerite is almost fused with the pleuron and its muscle pulls the whole pleuron (including the pleural wing process) downwards, closing the sternopleural hinge. The accessory indirect muscles are reduced but include a long, slender retractor extensoris (Rüschkamp, 1927) (*rem*, fig. 21B, C), opposing the pull of the basalar muscle, and a tergopleural muscle (*tpm*, fig. 21B, C), called by Stellwaag and Rüschkamp the 'relaxator alae'. The pleurosternal muscle is a flat band inserted by a long apodeme on the extreme posterior end of the pleural ridge at the point (*P*, fig. 20A) which forms the posterior pivot of the sternopleural hinge. Although it clearly reinforces the lateral wall of the thorax, it is situated so far from the wing that it can have only a small effect on the lateral stiffness of the articulation.

The relative size of the indirect wing depressor (dorsal longitudinal muscle) and the indirect wing levators (oblique dorsal, dorsoventral and tergocoxal muscles) varies considerably in different beetles. In *Acrocinus* (*Macropus*) *longimanus* (Cerambycidae) (Rüschkamp, 1927), the wet-weight ratio of the depressor to the levators is as 1 : 3·3; in *Dytiscus marginalis* (Korschelt, 1923) it appears to be even less. Rüschkamp discussed this inequality

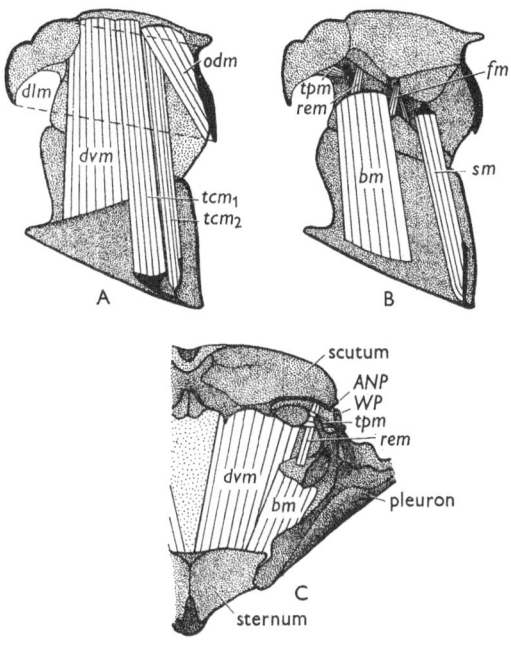

Fig. 21. The flight muscles of beetles. A, indirect muscles, B, direct and accessory indirect muscles of *Melolontha melolontha*, seen in internal view of the right metathorax. (Redrawn from Stellwaag, 1914.) C, anterior view of metathorax of *Dytiscus marginalis* (modified from Korschelt, 1923). *ANP*, anterior notal process; *bm*, basalar muscle; *dlm*, dorsal longitudinal muscle; *dvm*, dorsoventral muscle; *fm*, flexor muscle of 3rd axillary; *odm*, oblique dorsal muscle; *rem*, retractor extensoris muscle; *sm*, subalar muscle; *tcm*, tergocoxal indirect muscles; *tpm*, tergopleural muscle; *WP*, pleural wing process.

in size of the muscle groups producing the two directions of wing motion and decided that it could not be explained in terms of a residual depressor elasticity of the skeleton; he rightly concluded that the direct basalar and subalar muscles must assist in producing the downstroke, and showed that the ratio of indirect and direct depressor to levator muscles was 1 : 1. In that they use these two direct muscles as phasic muscles to assist in wing depression, the Coleoptera resemble the Orthoptera and Odonata, and are distinguished from the Diptera and Hymenoptera in which all the power for the wing strokes is provided by the indirect muscles.

The elastic properties of the metathorax are illustrated by the following observations on a coccinellid from which the elytra had been removed. When the insect was anaesthetized with CO_2 the unfolded wings took up a position below the horizontal, making an angle with one another of about $120°$; pressure on the notum produced a smooth movement of elevation to a vertical position. If the insect was anaesthetized with CCl_4, the wings rested at a relative angle of about $90°$ (more depressed) and pressure on the notum now produced only a sudden 'click' action. Pressure on the lateral parts of the notum had a markedly unsymmetrical influence on the wings so that one could be clicked up while the other remained unmoved. At times the wings would remain in the fully elevated position when the pressure was removed; they could never be brought into a steady intermediate position. This behaviour exactly resembles that described by Boettiger and Furshpan (1952) for the CCl_4-anaesthetized fly (p. 12). It is clear that here also there is a click mechanism which is being brought into action by muscles which contract tonically under the influence of CCl_4.

The downstroke of the beetle wing is produced by movement of both the anterior notal and pleural wing processes, neither of which can be thought of as a fixed pivot point. It is convenient to refer movements to the sternum, to which the anterior and posterior ends of the notum are joined through the mesothoracic and postnotal bridges. Contraction of the dorsal longitudinal muscle, occupying a median position in the segment, arches the notum on each side and raises the anterior notal process. Contraction of the basalar muscle pronates the wing and also, owing to the firm attachment of the basalar sclerite with the pleuron, moves the pleural wing process inwards and downwards under the anterior notal process by virtue of the hinge joint between pleuron and sternum; the subalar muscle depresses the wing in a similar manner but without pronation. A greater amount of wing depression is thus produced than would be possible by movement of one point alone. On the upstroke the pleuron is forced outwards and upwards again by the lowering of the anterior notal process through contraction of the dorsoventral muscle. In such a mechanism a local pleurosternal link would be ineffective as a means of increasing lateral stiffness, both on account of the hinge between the pleuron and the sternum and because the mean position about which the toggle action must operate is below the horizontal; the pleurosternal muscle is, in fact, inserted far back at the hinge pivot. The click action must be produced by simultaneous contraction of the dorsal longitudinal muscles forcing the anterior notal process outwards (as in *Sarcophaga*, p. 12) and of the basalar and subalar muscles forcing the pleural wing process inwards (as in Odonata, p. 23); the tergopleural muscle (*tpm*, fig. 21 B, C), which is exactly placed to increase the force between these two wing articulations, may assist and vary the click action.

OTHER ORDERS

More or less detailed accounts of the wing articulation and flight musculature have been published for a number of other insect orders. For completeness the main recent references are listed below, together with notes on any features which are of importance for an understanding of later chapters.

Orthoptera: Voss (1905); Carpentier (1923); Carbonell (1927); Crampton (1927); Snodgrass (1929); Solf (1931); Thomas (1952, 1953); Tiegs (1955). In the metathorax of blattids, of mantids and of *Gryllotalpa* the median tergal muscles are very ineffective in wing depression compared with the basalar and subalar muscles; in the mesothorax of blattids and mantids the oblique dorsal muscles are important wing depressors. Only in the gryllids and grasshoppers do the dorsal longitudinal muscles become important, forming in the Acrididae the main wing depressors (Tiegs, 1955).

Plecoptera: Wittig (1955).

Isoptera: Fuller (1925).

Odonata: Cremer (1934); Sargent (1937); Clark (1940); Chao (1953).

Hemiptera: Weber (1929, 1930); Tiegs (1955). In *Perkinsiella saccharicida* (Delphacidae) a dorsal longitudinal muscle of the first abdominal segment becomes an accessory depressor of the hind wing (Tiegs, 1955).

Neuroptera: Weber (1928); Maki (1936); Sundermeir (1940).

Mecoptera: Hasken (1939).

Lepidoptera: Weber (1928); Chadwick (1953); Nuesch (1953, 1954). A well-developed pleurosternal muscle is found in *Samia* (Chadwick, 1953), but in *Zygaena* and *Sphinx* the mesopleural and sternal apophyses are fused to produce rigid lateral bracing for the thorax (Weber, 1928).

Coleoptera: Stellwaag (1914); Korschelt (1923); Rüschkamp (1927).

Hymenoptera: Stellwaag (1910); Weber (1927a, b); Snodgrass (1910, 1925, 1942, 1956).

Diptera: Ritter (1912); Mihalyi (1935–6); Williams and Williams (1943); Zalockar (1947); Bonhag (1949); Sara and Smerdel (1953); Tiegs (1955). The subalar muscle is present only in the more primitive groups, e.g. Tipulidae; the oblique dorsal muscle is an important wing levator (Tiegs, 1955).

General works: Snodgrass (1909, 1927, 1935); Voss (1913); Maki (1938); Chadwick (1953).

CHAPTER 3

The Histology, Physiology and Biochemistry of Flight Muscle

THE existence in certain insects of an unusual type of striated muscle has been known, from histological evidence, for over a hundred years (von Siebold, 1848). It is only recently, however, that it has been demonstrated that also physiologically (Pringle, 1949) and biochemically (Watanabe and Williams, 1951) there is present here a tissue of exceptional interest.

HISTOLOGY

The muscles of insects differ from those of vertebrates in several ways. The fibres of which they are composed are always inserted at both ends directly into the cuticle of the exoskeleton, and lack the connective tissue envelope from which vertebrate tendons are derived. Where structures analogous to tendons are present, these are formed by the invagination of a region of the cuticle, and for distinction such structures are referred to as apodemes. Insect muscles are always striated, the striation being a feature of the myofibrils of which they are composed, and it is these myofibrils which, by their terminal tonofibrillae, achieve a direct transmission of their tensile force to the skeleton (Korschelt, 1938).

The leg and 'normal' trunk muscles are often termed 'tubular' muscles, since a central core of fluid sarcoplasm containing the string of nuclei has, owing to the absence of the myofibrillar proteins, a lower optical density than the rest of the fibre (figs. 22 A, 23 A). Around this core the myofibrils are arranged in a pattern of radial rows which is particularly evident after fixation. Fibres of well-developed leg muscles are usually found to have a diameter of 10–30 μ, and the cross-striation is remarkably similar to that of vertebrate muscle. The small sarcosomes are arranged between the myofibrils, either irregularly or else as a double row on each side of the Z-disc (Retzius, 1890; Ciaccio, 1940) (fig. 24A).

33

It is generally supposed that the muscles used in flight in the most primitive insects were the normal trunk and leg muscles of the thoracic segments, and in some Orthoptera (table 1; Blattidae) there is very little difference in the histology of the flight and leg muscles. The fibres are tubular and of small diameter, with myofibrils less than $0.5\,\mu$ across, again a comparable dimension to that found in vertebrate striated muscle. The flight muscles develop by a gradual process of fibre proliferation from muscles present in the nymph (Tiegs, 1955). These insects are,

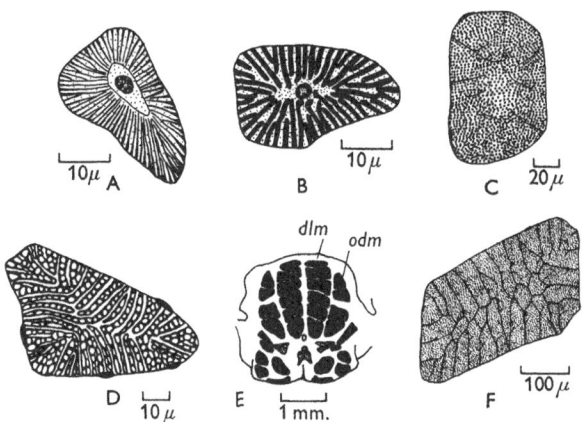

Fig. 22. Transverse sections of insect muscle. A, tubular leg muscle of *Vespa* (redrawn from Jordan, 1920); B, tubular flight muscle of *Libellula* (Odonata) (redrawn from Holmgren, 1910); C, close-packed flight muscle of *Chortoicetes terminifera* (Acrididae); D, lamellar flight muscle of *Cyclochila australasiae* (Cicadidae); E, transverse section of entire thorax of *Thelaira* sp. (Diptera, Tachinidae) showing only six fibres in each dorsal longitudinal muscle (*dlm*) and four in each oblique dorsal muscle (*odm*); F, fibrillar flight muscle of *Musca domestica* (C, D, E, F redrawn from Tiegs, 1955).

in general, weak fliers and there is probably no call for a metabolic output greater than that of the leg muscles.

From this condition it appears that there are three main types of histological evolution. In the Odonata (fig. 22 B) the tubular structure is preserved in all the flight muscles, with slightly larger myofibrils radially arranged in small-diameter fibres. The main difference between these muscles and a typical leg muscle is the presence of large sarcosomes regularly arranged between the myofibrils at the level of the *A*-band. The sarcosomes have been shown in dipteran muscle to be mitochondria, containing the enzymes for oxidative phosphorylation (Watanabe

TABLE 1. *Histological characteristics of indirect flight muscles*

Order	Family	Species	Muscle type	Fibre diameter (μ)	Fibril diameter (μ)	Sarcomere rest length (μ)	Sarcosome diameter (μ) and arrangement	Tracheation	Innervation	Development	Author
Orthoptera	Blattidae	*Blattella germanica*	Tubular	12–20	0.5	2.7	—	Surface	—	Fibre proliferation	Tiegs (1955)
	Gryllidae	*Acheta commodus* =*Gryllus servillei*	Irregular tubular	18–40	—	3.0	—	Surface	—	Fibre proliferation	Tiegs (1955)
	Tettigoniidae	*Acridopeza reticulata*	Close-packed	50	—	4.8	3.0: irregular	Some intracellular branches	End-organ	Fibre proliferation	Tiegs (1955)
	Acrididae	*Chortoicetes terminifera*	Close-packed	60–100	0.5–1.0	4.0	Large: regular at *A*	Intracellular network	—	Fibre proliferation	Tiegs (1955)
Odonata	Libellulidae	*Libellula* sp.	Tubular	20–25	0.7	3.0	1.7: regular at *A*	Intracellular network	—	From nymphal muscles†	Holmgren (1910); †Marcus (1920), Cremer (1934)
Hemiptera	Cicadidae	*Cyclochila australasiae*	Lamellar	60–90	<1	3.0	1.5: irregular	Intracellular network	End-organ: two fibres	Fibre cleavage	Tiegs (1955)
	Cercopidae	*Bathylus albicinctus*	Pseudo-fibrillar	20–50	>1	4.5	Small irregular	Surface	—	Fibre cleavage and myoblast incorporation	Tiegs (1955)
	Jassidae	*Erythroneura ix*	Pseudo-fibrillar	70	2.0	3.0	Regular	Intracellular network	End-organ: two fibres	Myoblast incorporation on rudiment	Tiegs (1955)
	Belostomidae	*Belostoma* sp.	Fibrillar	70	0.7–1.0	2.8	Regular	—	—	—	Edwards *et al.* (1954 *b, c*) Keilich (1918), *Ciaccio (1887)
Lepidoptera	Sphingidae	*Sphinx ligustri*	Close-packed	70	—	—	—	—	End-organ*	—	Keilich (1918) Edwards *et al.* (1954 *a*)
Coleoptera	Dytiscidae	*Dytiscus marginalis*	Fibrillar	90	3.0	2.5	3.0: regular	Intracellular	—	—	Jackson (1933)
	Hydrophilidae	*Hydrous piceus*	Fibrillar	70	2.2–5.4	—	—	—	—	—	Jordan (1920)
	Curculionidae	*Sitona lineatus*	Fibrillar	110–240	3.0–4.0	—	1.0–2.4	Intracellular	—	—	Morison (1928)
Hymenoptera	Vespidae	*Vespa* sp.	Fibrillar	175	3.0	3.0	—	—	—	—	Tiegs (1955)
	Apidae	*Apis mellifera* ♀	Fibrillar	170–200	2.4–3.0	2.5	Large	—	—	—	Tiegs (1955)
Diptera	Tipulidae	*Neoaratus hercules*	Fibrillar	120–170	3.5	3.0–4.0	Large	Intracellular branches	—	—	Tiegs (1955)
	Drosophilidae	*Drosophila melanogaster*	Fibrillar	100	1.6	—	Large	Intracellular network	Intracellular	Myoblast organization	Tiegs (1955)
	Tachinidae	*Rutilia potina*	Fibrillar	1800	2.0	3.0–4.0	Large	Intracellular network	Intracellular	—	Tiegs (1955)
	Muscidae	*Phormia regina*	Fibrillar	—	—	—	1.0–4.0: regular at *A*	Intracellular network	—	—	Watanabe and Williams (1951)

and Williams, 1951) and other enzymes to be discussed later. Their large size and regular arrangement in the flight muscles of Odonata can be correlated with the need for a high, sustained rate of metabolism to provide the energy for flight. The process of development in Odonata has been studied by Cremer (1934), who finds the muscle fibre cells already present in the 1½-year nymph, but with large nuclei and no myofibrils; these begin to appear in the 2-year nymph, but the nymphal muscle is never contractile until after emergence.

The second type of histological structure is found in the higher Orthoptera and Lepidoptera, and has been called 'close-packed' in table 1. The radial pattern is now absent (fig. 22 c) and the abundant nuclei lie just underneath the sarcolemma. The myofibrils are packed tightly into the fibres which often assume a polygonal shape in cross-section; they may be divided into blocks by intracellular tracheal branches which show a tendency to closed net formation. The myofibril diameter is still small, but, as in the Odonata, large, regularly arranged sarcosomes are found in muscle fibres of powerful fliers. Both these orders use the basalar and subalar muscles to assist in the downstroke of the wing, and there is little difference in histology between these and the indirect muscles. Development in Orthoptera is by fibre proliferation from muscles or muscle rudiments present in the nymph; those not used by the nymph naturally show a greater amount of fibre cleavage in the developmental stages (Tiegs, 1955).

The third type is fibrillar muscle, which reaches its highest specialization in the Coleoptera, Hymenoptera and Diptera. The entire cross-section of the fibre is here filled with fibrils and irregularly arranged nuclei, and the fibres are commonly of large diameter, reaching 1·8 mm. in the tachinid fly *Rutilia potina*. In the Diptera, which have been extensively studied by Tiegs (1955), there is in many families a remarkable economy of fibre number, the dorsal longitudinal muscle, for example, containing only six fibres on each side throughout the Muscidae, Anthomyiidae, Tachinidae and in *Drosophila* (fig. 22 E); in many of the nematocerous families, by contrast, there may be as many as 450 fibres (*Phellus glaucus*, Asilidae) only 150 μ thick. The giant fibres are cleaved into almost separate bundles of fibrils by an intracellular tracheal network (fig. 22 F), and in the higher Diptera it appears that even the motor nerve branches pass down these furrows to end deeply in the body of the fibre. Tiegs discusses at some length the status of the fibrils, into which (as has been known for a long time) the fibres readily split on gentle teasing even in the fresh state; he has evidence that these are not the myofibrillar units themselves but bundles of adherent myofibrils for which he revives the name of 'sarcostyles'. The composite nature of the sarcostyle may give an explanation of its large diameter, which is always greater than 1μ and may reach nearly 5μ; on the other hand, electron-microscope studies (Hodge, 1955) show no such subdivision. It is clear, in any case, that the fibrils or sarcostyles are the functional units of fibrillar muscle, since the large sarcosomes always lie between them (fig. 24 B).

It has from time to time been reported (bee, Keilich, 1918, Morison, 1928; *Drosophila*, Williams and Williams, 1943) that a sarcolemma is absent from some types of fibrillar muscle. Tiegs (1955) shows clearly that the boundary membrane of the muscle fibre is present in all cases, and suggests that the failure of some investigators to find it is due to their confusing some of the intracellular tracheal clefts with true cell boundaries. Excellent electron micrographs of sections of the sarcolemma in fibrillar muscle from Coleoptera, Diptera and Hymenoptera are given by Chapman (1954), from which it is clear that it is a double membrane normally closely united but diverging in places to contain tracheolar branches.

A particularly interesting series of histological types of flight muscle has been described by Tiegs (1955) from the Homoptera (table 1). The Cicadidae have flight muscles of a specific type (fig. 22 D), with the fibres arranged in small bundles in a nucleated sheath and fibrils in complex lamellae, but the structural dimensions are not significantly different from those of the close-packed muscle type. The tymbal sound-producing muscle (Pringle, 1954 a) has a similar appearance. In the Cercopidae and, even more clearly, in the Jassidae there are several features of resemblance to fibrillar muscle. The nuclei occur in the general body of the fibre as well as under the sarcolemma, and the fibrils are thick (always greater than 1μ; up to 5μ in *Eurymela* sp.); development takes place by the progressive incorporation of myoblast cells into the muscle rudiment, producing finally a syncytial condition in which there is an equal number of fibrils and nuclei. The innervation of these fibres is, however,

37

still by surface end-plates, with two nerve fibres of unequal size as in many leg muscles (Mangold, 1905). Tracheation is intracellular but not rich, and sarcosome density and arrangement is, as usual, correlated with activity. This type of muscle has been called 'pseudo-fibrillar' in table 1; Tiegs suggests that it occupies an intermediate position between the close-packed and true fibrillar types. The structural dimensions of the fibres and fibrils in *Belostoma* (Heteroptera) flight muscle (Edwards, Santos, Santos and Sawaya, 1954 *b*, *c*) suggest that it may fall into the same category.

DISTRIBUTION OF MUSCLE TYPES. In the Coleoptera, Hymenoptera and Diptera the fibrillar structure occurs only in muscles responsible for the high-frequency movement of the wings; the wing adjustor muscles always show the normal tubular pattern. The high density of sarcosomes in the fibrillar muscles often gives them a pink or yellowish colour owing to the concentration of cytochrome and other substances concerned with oxidative metabolism, and they are therefore easily recognized in dissections of fresh material. A histological distinction can thus be made in these orders between phasic and tonic flight muscles. In the Hymenoptera the only fibrillar muscles are the dorsal longitudinal and tergosternal indirect muscles of the mesothorax. In Diptera the dorsal longitudinal, oblique dorsal, tergosternal and tergocoxal indirect flight muscles are fibrillar,[1] but the tergotrochanteral muscle, which has the double role of indirect wing elevation and trochanteral extension, is tubular (Miller, 1950), as are also all the direct and accessory indirect muscles. Williams and Williams (1943), confirming the observations of Behrendt (1940), showed that the ability to jump in *Drosophila* is lost if the mesothoracic legs are removed and suggested that this is the function of the tergotrochanteral muscle. Boettiger and Furshpan (1952), however, maintain that this muscle is important at the start of flight, producing the initial upstroke after the slow down movement which results from tetanic contraction of both sets of indirect muscles (p. 15). It seems possible that this is a true case of a double function, since a jump into the air is advantageous at the beginning of flight. A list of dipteran families possessing the tergotrochanteral muscle is given by Tiegs (1955) and Smart (1957); it

[1] The direct subalar muscle, found only in Nematocera, is also fibrillar (Smart, J.; unpublished).

38

is absent in Nematocera and in Bombyliidae, Asilidae and Stratiomyiidae. The main vibrator muscle of the haltere is also fibrillar.

In the Coleoptera the fibrillar muscles have a yellowish rather than a pinkish coloration. They are (fig. 21) the dorsal longitudinal, oblique dorsal, tergosternal and first tergocoxal indirect muscles of the metathorax, and also the direct basalar and subalar. The second tergocoxal (tcm_2, fig. 21 A) is not yellowish in colour in *Dytiscus marginalis*; it is inserted far out on the scutum and is an effective wing levator. Comparative studies of this muscle in different beetles have not been made, but it is possible that it functions as a starter, as in Diptera. The remaining metathoracic muscles, and all those in the mesothorax concerned with folding and unfolding the elytra, are of normal histological structure.

In the mesothorax of Jassidae and Cercopidae (Homoptera) the dorsal longitudinal, oblique dorsal, tergocoxal, basalar and subalar muscles are pseudo-fibrillar in structure and yellowish in coloration, with several tergal leg muscles normal (Tiegs, 1955); the metathoracic flight musculature of these families is reduced. In *Perkinsiella saccharicida* (Delphacidae) there is no tergocoxal flight muscle in the mesothorax, and in the metathorax the stout, fibrillar dorsal longitudinal muscle is balanced mainly by an abdominal dorsal longitudinal muscle which is pseudo-fibrillar and has become an accessory wing-levator. Here again the correlation between histology and function is clearly established.

In insects with muscles of the close-packed type it is more difficult to use histological criteria to decide which muscles are used phasically during flight; for example (Tiegs, 1955) has a photomicrograph of a leg muscle of *Acridopeza reticulata* (Orthoptera, Tettigoniidae) showing the same type of close-packed appearance in cross-section as in the flight muscles. The amount of intracellular tracheation is a better test; on this basis, Tiegs includes in the phasic flight muscles of the half-mesothorax of the Acrididae the dorsal longitudinal muscle, two tergosternals, three tergocoxals, a tergotrochanteral, two basalars and the subalar, and in the metathorax all these and one more basalar. An indication that not all of these are used in flight comes from comparative studies of flying and flightless forms. Thomas (1952, 1953), comparing normal males and apterous females of

Lamarckiana sp. and macropterous and brachypterous females of *Chorthippus parallelus*, found that the flightless condition is associated with reduction to fibrous strands of the dorsal longitudinal, tergosternal and first basalar muscles, with some reduction in size of the second basalar and the subalar. On the other hand, the oblique dorsal muscle (which is small in the Acrididae) and the tergocoxal muscles were not reduced. Ewer (1954*b*) found a very similar state of affairs in *Zonocerus elegans*. Indirect evidence of this sort is not conclusive against a role in flight of the latter group, but it is noteworthy that similar studies of flightless beetles (Jackson, 1933, 1952) have shown a good functional correlation; there is degeneration of the fibrillar basalar and subalar as well as of the indirect muscles, but the lateral tergocoxal muscle (of *Anacaena globulus*, Hydrophilidae), which is of the tubular type, is present and is used for spreading the wings.

STRIATION. The figures for sarcomere rest length in various species (table 1) show no good correlation with histological type or with the speed of contraction as judged by wing-beat frequency. On the other hand, within a given species, Edwards *et al.* (1954*c*) have found that fibrillar muscles have a consistently shorter sarcomere rest length than the tubular leg muscles. It has often been claimed (Jasper and Pezard, 1934) that faster-acting muscles have a shorter striation period, but the wide range found in insect flight muscles hardly supports this view. This is, however, a particularly difficult measurement to make, and too much attention should not be paid to the minor variations of this parameter in table 1.

Insect muscle was a favourite material among the older histologists for the study of striation pattern and its changes during contraction. Jordan (1933) gives the more important references. Fig. 23 shows the appearance of typical tubular and fibrillar muscles in lightly stained preparations viewed with the light microscope; these and later drawings are lettered according to the currently agreed terminology. Because of optical difficulties at the limit of resolution much of the early work on changes in the pattern of striation during stretch and contraction is now held to be unreliable, and recent studies have concentrated on the use of phase-contrast, interference, polarizing and electron microscopes. Differences between vertebrate, insect tubular, and insect fibrillar muscle appear to be quantitative rather than qualitative; no new features of the pattern of striation have been established for insect muscle, but there is variation in the relative lengths of the anisotropic (*A*) and isotropic (*I*) bands and in the prominence of the *H*-region and of the contraction bands.

Fibrillar muscle has been studied by modern methods by Edwards *et al.* (1954*a*, *b*, *c*), Chapman (1954), Philpott and Szent-Györgyi (1955), Hodge (1955, 1956), Edwards and Ruska (1955) and Hanson (1956*a*, *b*). Isolated fibrils can be obtained with or without glycerol extraction and preserve the pattern of striation. In *Calliphora* such glycerol-extracted fibrils show a

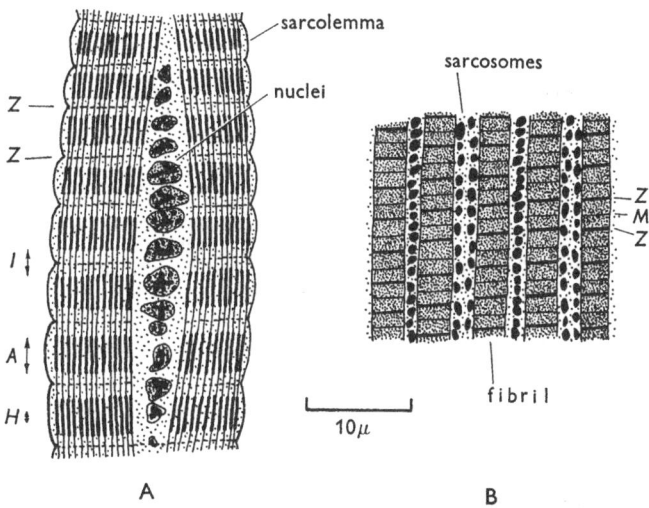

A   B

Fig. 23. Longitudinal sections of *Vespa* muscle. A, tubular leg muscle fibre; B, portion of fibrillar flight muscle fibre. The bands are lettered according to the currently agreed terminology. (Redrawn to the same scale from Jordan, 1920.)

variation in sarcomere length from 3·2 to 3·7 μ (Hanson, 1956*b*); the longer fibrils show an *I*-band 0·7 μ long, those of intermediate length no *I*-band, and the shortest fibrils contraction bands on each side of the *Z*- and *M*-lines. This last appearance is that usually shown by fixed and stained fibrils (fig. 23 B). Hanson (1956*a*) could change the isolated fibrils from one condition to another by treatment with adenosine triphosphate (ATP) or muscle extract, and she emphasizes the small length change which is involved (about 6 %). If passively stretched by more than this small amount the fibrils form regions of very low density on one or both sides of the *Z*-line; these irreversibly stretched regions have been incorrectly described as *I*-bands in some earlier papers by authors using the light microscope.

The electron micrographs of Edwards *et al.* (1954 *a, b, c*) establish the shortness of the *I*-band of fibrils from the fibrillar muscles of *Hydrous* (*Hydrophilus*) *piceus* (Coleoptera) (0·39μ in a total sarcomere length of 2·47μ = 16 %) and *Belostoma* sp. (Heteroptera) (0·52μ in a total of 2·78μ = 18 %), and show also the tendency to form contraction bands and to extend on stretch by separation on each side of the *Z*-line. Philpott and Szent-Györgyi (1955) give some excellent pictures of the internal organization of the fibrils in the housefly and bumble-bee, but their discussion shows a lack of understanding of the normal method of functioning.

There would seem to be a progressive reduction in the relative rest length of the *I*-band in the more specialized flight muscles.

Rest length of I-band expressed as percentage of sarcomere length

Vespa (Jordan, 1920)	tubular leg muscle	50
Libellula (Holmgren, 1910)	tubular flight muscle	30
Schistocerca (Weis-Fogh, 1956 *c*)	close-packed flight muscle	21
Calliphora (Hanson, 1956 *b*)	fibrillar flight muscle	0–20

This can probably be correlated with the more nearly isometric conditions of operation.

Ultra-thin sectioning of muscle has been used by a number of authors to study the pattern of arrangement of protein filaments within the myofibril, and there is at present a difference of opinion in the interpretation of the electron micrographs between Hodge (1955, 1956) and Hanson and Huxley (1955); different theories about the molecular mechanism of muscular contraction result from the two interpretations. Until this disagreement is resolved it is too early to discuss any possible fundamental difference in the fine structure of the fibril between insect fibrillar and mammalian muscle.

BIOCHEMISTRY

In the striated muscle fibres of insects, as of other animals, there is a definite localization of biochemical substances in different parts of the cell, and since fibrils, sarcosomes and sarcoplasm can be obtained by centrifugation in an almost pure preparation, much of our knowledge of intracellular biochemical specialization has been obtained from these tissues.

The structural proteins of muscle (mainly myosin and actin) are located in the fibrils. It is now widely accepted that the chemical energy for the contractile activity of these proteins is supplied in the form of energy-rich organic phosphate compounds synthesized elsewhere in the cell through the operation of the metabolic enzymes. Watanabe and Williams (1951) were the first to show that the complex of respiratory enzymes in insect fibrillar muscle is located exclusively in the sarcosomes, though Keilin (1925) had long before identified cytochrome in many flight muscles and emphasized the correlation between its

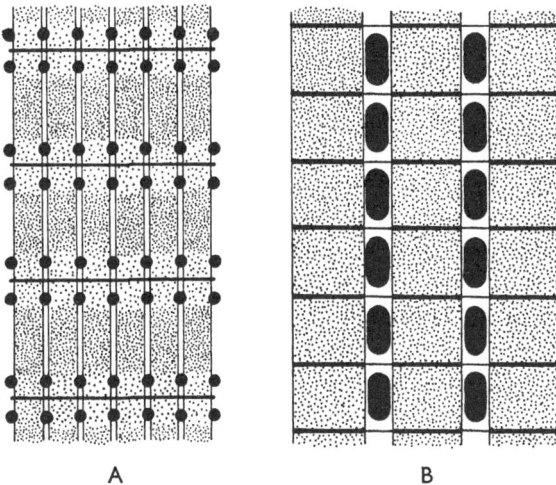

A   B

Fig. 24. Diagrams showing the arrangement of sarcosomes. A, leg muscle; B, flight muscle, of *Hydrous piceus* (Coleoptera). (Redrawn from Ciaccio, 1940.)

concentration and muscular activity. Since then all studies of the enzyme biochemistry of insect muscle have re-emphasized the importance of the sarcosomes, which provide one of the most active complexes of metabolic enzymes known.

Ciaccio (1940), Cleland and Slater (1953) and Tiegs (1955), in independent researches, have confirmed the distinction made by Holmgren (1910) between muscles with small sarcosomes arranged either irregularly or at the level of the *I*-band and those with large sarcosomes regularly disposed opposite *A* (fig. 24). This distinction cuts across the classification of flight muscles by fibril pattern (table 1) and is clearly correlated with activity. Cleland and Slater (1953) consider that, in the former type, metabolism is often largely anaerobic during intermittent activity. The enzymes for anaerobic glycolysis are found dispersed in the sarcoplasm (Saktor, 1955); complete oxidation of the metabolic substrates demands a high concentration of oxidative enzymes and carriers which are located exclusively in the sarcosomes.

The nature of the substrate used during flight has been investigated in *Schistocerca* (Krogh and Weis-Fogh, 1951; Weis-Fogh, 1952), in *Drosophila* (Williams, Barnes and Sawyer, 1943; Chadwick, 1947; Wigglesworth, 1949) and in a number of other insects. During flight in *Drosophila* the respiratory quotient rises to 1·0, indicating a utilization of carbohydrates; in *Schistocerca* it falls from 0·82 to 0·75, showing that fat is being mobilized. Carbohydrate is also used in the blow-fly *Lucilia sericata* (Davis and Fraenkel, 1940; Williams *et al.*, 1943) and the bee *Apis mellifera* (Jongbloed and Wiersma, 1935; Beutler, 1937) and fat possibly in the leafhopper *Eutettix* (Fulton and Romney, 1940) and the butterfly *Danaus* (Beall, 1948). In *Drosophila* and other flies, glycogen in the flight muscles and elsewhere in the body provides the immediate fuel supply (Wigglesworth, 1949). In *Apis* glycogen reserves are small and the fuel appears to be glucose, of which the normal concentration in the blood is 2·6 %; below 1 % the bee is unable to fly (Beutler, 1937). In insects which use mainly fat, it is not clear whether the muscles themselves can metabolize fatty acids or have to be supplied with an oxidative breakdown product such as acetate by enhanced catabolism in the fat body. Studies on the oxygen consumption of flight muscle extracts in *Periplaneta* (Barron and Tahmisian, 1948) and *Locusta* (Rees, 1954) have shown an inability to metabolize, respectively, butyric and octanoic acids, but acetate can be used, presumably by means of coenzyme-A and the citric acid cycle, the enzymes for which are present in the sarcosomes. A low fatty-acid oxidase activity compared to rat liver was also found by McShan, Kramer and Schlegel (1954) in the thoracic muscles of the woodroach, *Leucophaea maderae*. Weis-Fogh (1952) points out that even in the locust carbohydrate is consumed at the start of flight, and correlates this with the greater ease of mobilization of glycogen rather than fat, which is a more suitable fuel for long-distance, steady fliers.

The most complete analysis of enzymes and substrate potentialities of flight muscle has been made by Saktor (1953*b*, 1955) on extracts of thoraces of *Musca domestica;* this has the highly specialized fibrillar muscle, with large sarcosomes. Saktor's flies were starved before use, and the rate of oxygen uptake of his extracts corresponds approximately to that of the insect at rest; he is not, therefore, on very safe ground in drawing conclusions from his results about the properties of active muscle. Nevertheless, it is interesting that by recombining sarcoplasm and particulate suspension (fibrils *plus* sarcosomes) he has been able to obtain a metabolic activity about half that of the original homogenate and to separate the glycolytic enzymes in the sarcoplasm from the enzymes of the citric acid cycle in the sarcosomes. In addition to sugars and hexose phosphate, certain particular amino acids could be oxidized by the whole extract and also acetate and glycerol, but his figures show only low activity for the latter substances. A peculiarity is the absence of lactic dehydrogenase, and evidence from other insects (cockroach, Barron and Tahmisian, 1948; Humphrey, 1949; locust, Humphrey and Siggins, 1949) also suggests that pyruvic or phosphopyruvic acid and not lactic acid is the end-product of glycolysis in insects. Saktor's sarcosome preparations readily utilized α-glycerophosphate, and it is noteworthy that in a recent study of a number of insect muscle extracts Zebe (1957) has found the most active enzyme to be a glycerophosphate dehydrogenase; he also confirms the low lactic dehydrogenase activity but is unable to find differences in muscle enzymes as between species metabolizing fat or carbohydrate.

A comparison of the oxygen consumption and succinic dehydrogenase activity between teased leg and flight muscles has been made by Pérez-González and Edwards (1954) for *Periplaneta americana*, *Schistocerca infumata*, *Belostoma* spp. and *Hydrophilus ater*. The cockroach showed little difference in either ratio, and in the locust the dehydrogenase activity of flight muscle was only three times that of leg muscle. *Belostoma* and *Hydrophilus* flight muscles, however, which are fibrillar, were 15–20 times more active than leg muscle, though the absolute values were no higher than for the orthopteran species. Here again the normal rate of oxygen uptake of active flight muscle was not nearly approached.

The very high total metabolic rate of intact flight muscles is stressed by Weis-Fogh (1952) and Chadwick (1953). The indirect muscles of some Diptera and Hymenoptera reach values of 2000 kcal./kg./hr., nearly twice that calculated for humming

45

birds and ten times the maximum for human heart muscle. The increase in rate of oxygen consumption during flight may be as much as fifty times (*Apis*, Jongbloed and Wiersma, 1934; *Schistocerca*, Krogh and Weis-Fogh, 1951). In *Drosophila* an increase of over twenty times is achieved without appreciable oxygen debt (Chadwick and Gilmour, 1940), but in *Schistocerca* normal respiratory rates are not re-established for 1–2 hr. after prolonged flight. Both this difference and the different fuel consumed are consistent with a difference in mechanism as between non-fibrillar and fibrillar muscle, but the number of examples is not yet sufficient to justify a definite correlation. There is electron-microscope evidence (Philpott and Szent-Györgyi, 1955) that the sarcosomes of fibrillar muscle have a more open internal double-membrane structure than those of tubular leg muscle; this could be connected with their more intense enzymic activity.

In fibrillar muscle and in many other types the sarcosomes are regularly aligned with the striations of the fibrils (fig. 24). The only evidence of a direct functional association is the observation of Hanson (1952) that the ATP-induced contraction of isolated fibrils of *Dytiscus* fibrillar muscle occurs only when sarcosomes adhere to the fibrils. There is, however, much evidence for the presence in the sarcosomes of enzyme systems concerned with oxidative phosphorylation. After an earlier failure (1953*a*) due to incorrect conditions, Saktor (1954) was able to demonstrate a P:O ratio of up to 1·6 using α-ketoglutarate as substrate and glucose *plus* hexokinase as a phosphate-acceptor system. He was unable to show the phosphorylation of arginine which, since Baldwin and Needham's (1934) identification of arginine phosphate in insect muscle, has been held to be the main 'phosphagen' energy-rich phosphate reserve in this as in other invertebrate tissues. Lewis and Slater (1953) found similar P:O ratios in *Calliphora* sarcosome preparations and pointed out that this is lower than the value for vertebrate heart muscle. Although evidence such as the similarity in the effects of temperature (Saktor and Sanborn, 1956) all points to a close link between oxidative metabolism and phosphorylation in insect muscle, it seems probable that the optimum conditions have still not been reproduced in these sarcosome preparations, and that there are some undiscovered differences from the vertebrate systems. For example, Slater and Lewis (1953) claim that dinitrophenol increases phosphorylation in *Calliphora* thorax extracts, though this is denied by Saktor and Cochran (1956) for *Musca*.

The adenosine triphosphatase enzymes of insect flight muscle have been studied by Saktor (1953a), Saktor, Thomas, Moser and Bloch (1953), Gilmour (1953) and Gilmour and Calaby (1953a, b), and found to be very active in several fractions. Saktor was able to distinguish between the enzymes of fibrils, sarcosomes and sarcoplasm by different reactions to activators and inhibitors. A variety of differences have been found by these authors between flight muscles and leg muscles, but the studies do not yet give any consistent picture of the mechanism of energy utilization.

The conclusion which may be drawn from the work which has so far been done on the biochemistry of insect flight muscle is that in its general organization it is similar to vertebrate muscle, but with certain differences whose significance cannot at present be gauged. Although under certain conditions the fibrils will contract when treated with ATP, it is not certainly established that this substance is responsible for the transfer of energy from the sarcosomes to the fibrils; the nature of the energy-rich phosphate reserve (if any) is not clear. There is a correlation between the presence of large sarcosomes containing the oxidative enzyme complex and the ability to perform prolonged flights, and there is a mechanism in the sarcoplasm for the initial breakdown of metabolic catabolites, which may, in some cases, be used to supply extra energy, with an oxygen debt which is subsequently repaid. Fat stores can be mobilized sufficiently rapidly to prolong flight in some insects, though the biochemical pathway is unknown; others, particularly those with fibrillar flight muscles, have to use glycogen or glucose. The integration of all the enzyme systems into a physiological mechanism undoubtedly involves spatial relationships which are disturbed by extracting procedures, which tend to produce a condition in which hydrolysis predominates over synthesis to such an extent as to make the preparations highly abnormal. It should always be remembered that no extract has yet achieved one-hundredth of the metabolic rate of the flying insect.

PHYSIOLOGY

Of the various types of flight muscle which have been characterized histologically, only a few have been used for physiological investigations. The small size of many insects and the construction of the thorax often make it difficult to obtain an isolated muscle preparation, and in even fewer cases can the motor nerve supply also be dissected free from the body of the insect. This practical restriction to experiment has been particularly thwarting in the case of fibrillar muscle, and much of the progress made recently in understanding the mode of action of this tissue has been gained by indirect methods rather than by the usual direct attack of the neurophysiologist.

LEG MUSCLES. It is necessary to start this section with an account of the physiology of insect leg muscle, which is now reasonably well understood through the work of Hoyle (1955 a, b and earlier papers). Although the metathoracic extensor tibialis muscle of *Locusta migratoria*, with which he has worked, is in many ways very specialized for jumping, the three motor nerve fibres which innervate it produce effects which appear to cover all the cases of motor innervation described earlier for other muscles; these include various distal leg muscles of *Periplaneta americana* (Pringle, 1939), the tergal remotor of the coxa of *Periplaneta* (Roeder and Weiant, 1950), and various leg muscles of *Decticus verrucivorus* (Solf, 1931) and *Dytiscus* (Kraemer, 1932). The three motor nerve fibres to the locust jumping muscle are named by Hoyle (1955) F (fast), S_1 and S_2 (slow), and their effect on the muscle is shown in table 2. The arrival of an impulse in the F nerve sets up a localized depolarization of the muscle-fibre membranes which is followed about 10 msec. later by a twitch contraction lasting about 100 msec. (fig. 25A); there is no increase in the size of the localized 'end-plate potential' (e.p.p.) with repetitive stimulation and no propagation along the muscle-fibre surface membrane as in vertebrate muscle. Smooth tetanus results from repetitive stimulation at 25 per sec. This is a typical result for a fast fibre in insects. More rapid mechanical effects have been reported (Roeder and Weiant (1950) obtained a twitch lasting only 15 msec. in the tergal remotor muscle of *Periplaneta*); but it is unusual for any sign of the individual twitches to persist above stimulus frequencies of 30 per sec.

Of the two slow fibres, S_1 has endings of two types on different muscle fibres. The S_{1b} endings produce e.p.p.'s similar to but smaller than those of the F fibre, but show facilitation with repetitive stimulation (fig. 25B); the S_{1a} endings produce only

TABLE 2. *Electrical and mechanical response of a locust leg muscle to stimulation of its motor nerve fibres* (from Hoyle, 1955 *b*)

Type of nerve ending	F	S_{1a}	S_{1b}	S_2
Approx. % of muscle fibres innervated	100	20	10	
Electrical activity of muscle fibres:				
Single impulses	Large e.p.p.	Slow depolar-ization	Small e.p.p.	Slow hyper-polarization
Repetitive excita-tion	Series of e.p.p.'s	Summation of depolarization	Facilitation of e.p.p.s	Summation of hyper-polariza-tions
Mechanical activity of muscle fibres:				
Single impulses	Large twitch	Nil	Small twitch	Nil
Repetitive excita-tion	Tetanus	Tonic con-traction	Weak tetanus	Nil
Normal function	Jumping	Slow move-ments and maintained tension	Quicker move-ments	Preparatory (?)

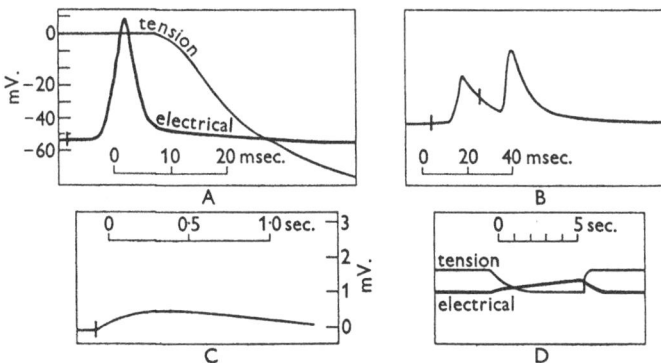

Fig. 25. Types of membrane response in insect leg muscle on stimulation of the motor nerve fibres, as recorded intracellularly by Hoyle (1955 *b*) from the extensor tibialis of *Locusta migratoria*. A, fast (*F*) fibre; B, slow (S_{1b}) fibre; C, slow (S_{1a}) fibre; D, slow (S_{1a}) fibre, repetitive stimulation at 50 per sec. The vertical strokes indicate the instants of stimulation.

slow depolarizations which, however, summate to give rise to large membrane-potential changes at frequencies of 15–150 per sec. (fig. 25 c, d). Both types of ending produce slow tension changes in the fibres with repetitive stimulation, and little or no twitch tension with single impulses. The S_2 fibre has been reported only from this preparation and is not fully understood;

it produces no mechanical activity, but hyperpolarizes the muscle-fibre membranes when these are depolarized below their normal level. There is no good evidence for the presence in insects of inhibitory motor nerve fibres such as have been found in Crustacea.

NON-FIBRILLAR FLIGHT MUSCLE. Early work on the physiology of flight muscles was mainly concerned with the problem of the high frequency of discrete twitches necessary to produce the wing beats. Heidermanns (1931), working with muscles from *Aeshna coerulea*, was the first to point out the importance of the nearly isometric conditions under which these muscles operate in the insect body. When isolated and studied isotonically with light loading the muscle showed fusion of contractions at about 20 stimuli per sec., whereas the normal wing-beat frequency is 25 per sec.; even with optimum loading (which according to Cremer (1934) is about 750 times the muscle weight) he was forced to the conclusion that only half the possible twitch tension could be generated in the time available, and that the muscles must be working in partial tetanus. A similar conclusion was reached by Tiegs (1955) for the hawk-moth *Hippotion*, whose muscles go into partial tetanus when stimulated at 30 per sec. although the wing-beat frequency is 57 per sec. A possible resolution of the difficulty is suggested by the results of Buchthal, Weis-Fogh and Rosenfalck (1957), who find in *Schistocerca* that, while experiments with the isolated metathoracic dorsal longitudinal muscle suggest a similar state of affairs, if allowance is made for the 8° C. rise in temperature known to occur in the locust thorax during flight, the twitch duration is reduced sufficiently to avoid any tetanus at the normal frequency of wing beat; twitch duration always has a high temperature coefficient (Hill, 1951). The thoracic temperature during flight is above that of the environment in many insects (Sotavalta, 1954a) and may be raised before flight by a period of wing-whirring or increased respiratory movements during which it increases to what appears to be a critical level (Dotterweich, 1928; Krogh and Zeuthen, 1941). It is not, however, excluded that some unusual physiological properties may be present to assist re-extension (see p. 64). The problem is complicated by elastic forces in the flight system which may be much larger than has hitherto been supposed. In *Schistocerca* a large elasticity in the muscle can act sufficiently rapidly to produce significant assistance to the twitch even at muscle lengths found in the body (Buchthal *et al.* 1957).

The much higher frequencies of wing beat found in orders with fibrillar muscles present a different problem which will be discussed shortly, and it is convenient first to complete the review of work on 'normal' flight muscle. It seems that all the known examples of motor nerve supply to non-fibrillar flight muscle come in the class of fast fibres; there is no physiological evidence to support Tiegs's (1955) histological report of a double innervation in certain cases. Intracellular recording from flight muscles of *Locusta migratoria* by Hagiwara (1953) and Hagiwara and Watanabe (1954) showed large, non-propagating e.p.p.'s at each impulse with no facilitation. Some unusual rhythmic responses described from locust flight muscle by Voskresenkaya (1947) were reinvestigated by Ewer and Ripley (1953), who found them to be due mainly to after-discharge of motor-nerve impulses from the ganglion; the metathoracic tergosternal muscle of *Locusta migratoria* used by them gives a normal fast-fibre response, and the only unusual feature found was a prolonged nerve refractory period which was liable to produce intermittent excitation at near-threshold stimulus intensities.

The physiology of the metathoracic dorsal longitudinal muscle of *Schistocerca* in tetanus has been described by Weis-Fogh (1956*c*). The author points out that this muscle is never called upon to perform tetanic contractions in the living insect, and that it is, in fact, incapable of contracting isometrically without internal injury at the normal temperature of 25° C.; he found, however, that at 11° C. it could be made to contract for ½ sec. with repeatable results, although some fatigue was already apparent even after this short interval of time. His results are summarized in fig. 26. The passive muscle is relatively inextensible, tension rising steeply at greater than rest lengths; if this passive tension is subtracted, the isometric tension is maximal at about rest length, as is usual for striated muscles. The tetanic force at 25° C. is three times that at 11° C.; the temperature coefficient of passive tension is several times greater than for frog muscle (2–3 % per 10° C. compared to 0·5 %). If lightly loaded and allowed to shorten below about 50 % of rest length the muscle goes into the δ-state (Ramsey and Street, 1940) and its performance is then changed.

The passive elasticity of various insect muscles was studied in detail by Buchthal and Weis-Fogh (1956), and compared to that of frog muscle. In the frog the coefficient of passive elasticity increases rapidly as the muscle is stretched, much of the resistance being produced by the sarcolemma (Ramsey and Street, 1940). In *Schistocerca* flight muscle the sarcolemma contributes only one-third of the passive elasticity and less in locust abdominal muscle, while in the fibrillar flight muscle of *Hydrophilus* its contribution is negligible; apart from an initial portion

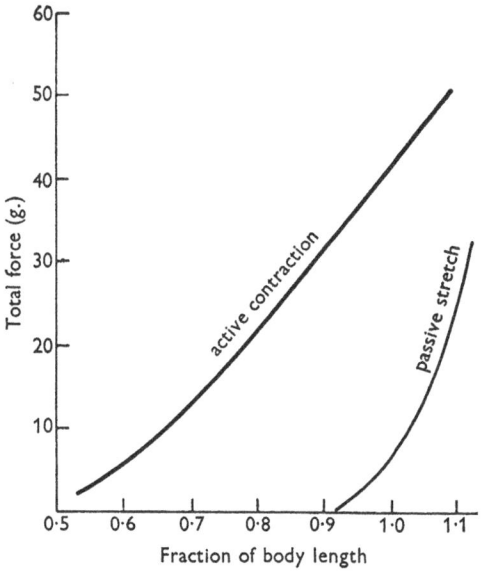

Fig. 26. Tension/length curves for active and passive metathoracic dorsal longitudinal flight muscle of *Schistocerca gregaria*. The active contraction curve includes points measured in isotonic and isometric experiments. (Redrawn from Weis-Fogh, 1956 c.)

probably due to slack fibres, the curve of tension/length is nearly linear. The authors conclude that the resistance to stretch must either be due to an elastic component of the contractile system, or else there must be parallel passive-elastic elements inside the fibre. Since there is no histological evidence for any fibrous structure in insect muscles other than the striated fibrils the former alternative seems more likely to be correct. If so, this passive elastic property of the fibrils represents a significant specialization of insect flight muscle with an important influence on the dynamics of the flight machine (Buchthal *et al.* 1957).

FIBRILLAR MUSCLE. The unusual physiological properties of fibrillar muscle were first pointed out by Pringle (1949) in relation to the flight and haltere muscles of *Calliphora*. It was already known from the work of von Buddenbrock (1919), Roch (1922), Sellke (1936) and Sotavalta (1947) that reduction of the wing area and weight by cutting increases the wing-beat frequency in Diptera; Pringle (1949) produced evidence that this is not a reflex effect from wing or muscle sense organs on a centrally determined nervous rhythm, but a direct influence of loading on the frequency of a myogenic rhythm. Electrical records from the indirect muscles in the thorax of a flying fly showed normal e.p.p.s of the fast nerve-fibre type, but at a frequency much less than that of the wing beats. This has been verified by Roeder (1951) for some other Diptera (*Lucilia, Eristalis*) and *Vespa* (Hymenoptera), and by Boettiger and Baranowski (unpublished) for representatives of these orders and the Membracidae, Jassidae and Coreidae (Hemiptera) (fig. 27). The hypothesis was put forward that in these muscles the contraction at each wing beat is not a normal twitch initiated by the arrival of an impulse in the motor nerve, but that the motor-nerve impulses, which produce in the muscles a normal type of electrical spike, bring the myofibrils into a state of excitation in which they contract actively only when stretched. In the flight system the two antagonistic sets of indirect muscles influence each other to contract, whereas in the haltere system (then believed to involve only one muscle with an elastic return) a mechanical resonance maintains the cycle. The failure to obtain any mechanical response from stimulation of the indirect muscles of *Calliphora* exposed by bisection of the thorax was explained by supposing that the disturbance to the mechanical system of the thorax had destroyed this mutual influence.

Independently of this work Boettiger and Furshpan (1950) had been studying the articulation of the Dipteran wing; Boettiger (1951) reported the results of electrical stimulation of an intact, flying fly. Control of the muscles could be taken over by the applied stimuli which produced large, all-or-none potentials, but the much higher frequency of wing beats continued unchanged. If a part of the notum was removed, the dorsal

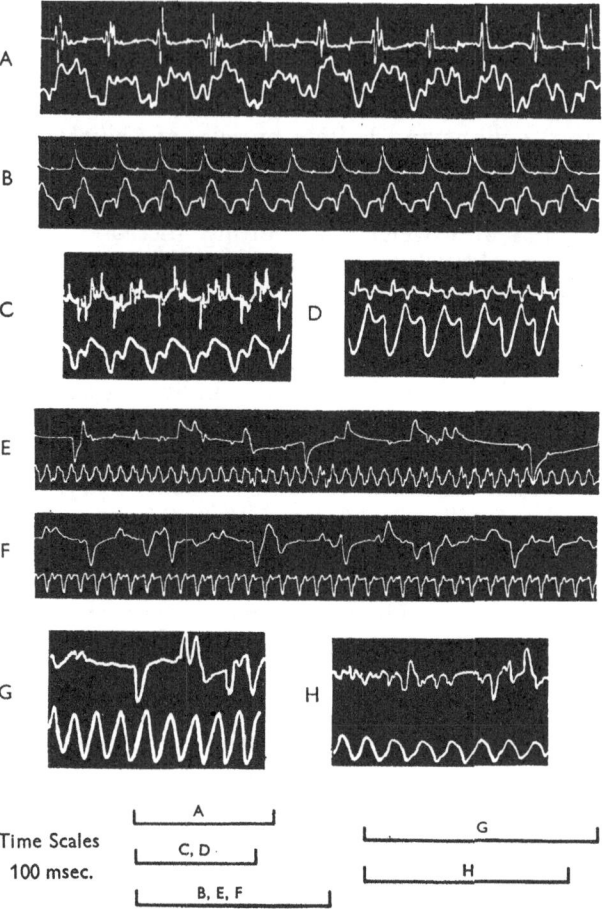

Fig. 27. Electrical (upper trace) and mechanical (lower trace) records from the thorax of various flying insects. A–D, synchronous (1:1) type; E–H, asynchronous (myogenic rhythm) type. A, *Periplaneta americana*; B, *Agrotis* sp. (Lepidoptera); C, a sphingid moth; D, a cicada; E, *Lucilia* sp. (Diptera); F, *Vespa* sp. (Hymenoptera); G, a membracid (Homoptera); H, a cicadellid (Homoptera, Jassidae). (A, B, E, F from Roeder, 1951; C, D, G, H from Baranowski, unpublished, retouched.)

longitudinal muscles could be seen to go into a smooth tetanus at a frequency of 10–20 stimuli per sec., and Boettiger concluded that the flight mechanism could be maintained in action by appropriate stimulation of the ganglion which resulted in simultaneous tetanic contraction of both antagonistic indirect muscle groups, the oscillation to produce the wing motion resulting from the click mechanism previously described (pp. 12–15).

Boettiger and McCann (1953) later verified with intracellular electrodes that the membrane properties of dipteran fibrillar muscle show no unusual features except a large, prolonged negative after-potential following the active spike of 90–100 mV. (resting potential 60–70 mV.). McEnroe (1953) verified the tetanic state of a fragment of the dorsoventral fibrillar muscle freed, in a tethered flying *Sarcophaga*, from one end of its normal attachments.

Further elucidation of the nature of the mechanical cycle of events in fibrillar muscle was obtained by Pringle (1954 *a*, *b*) working not with a flight muscle but with the tymbal (sound-producing) muscle of cicadas of the genus *Platypleura*. This muscle, when excited, contracts almost isometrically against the

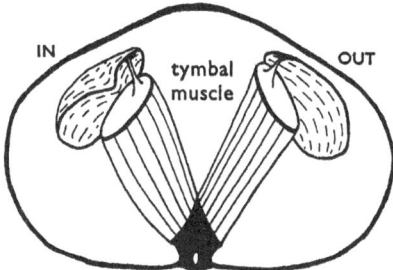

Fig. 28. Diagram to show the buckling of the cicada tymbal under the pull of the tymbal muscle (from Pringle, 1956).

elastic resistance of the tymbal cuticle until, at a critical tension controlled by a tonic tensor muscle, the tymbal suddenly buckles (fig. 28) and emits a pulse of sound. The material has the advantage over flight muscle that an isolated nerve-muscle-tymbal preparation can be made without impairing normal functioning. A single impulse in the single motor nerve fibre to the muscle elicits four sound pulses (*Platypleura capitata* at 30° C.), and a stimulus frequency of 50 per sec. is adequate to set up a myogenic rhythm of activity at about 320 per sec. (fig. 29 A, B). The electrical record from the muscle shows normal fast-fibre spikes synchronous with the stimuli. Since the tymbal muscle when connected to an isometric lever gives normal twitch responses to each stimulus and almost smooth tetanus at 50 per sec. (fig. 29 C, D, E), Pringle concluded that the click action of the tymbal was an essential feature for the myogenic rhythm, and that the important phenomenon for the maintenance of the

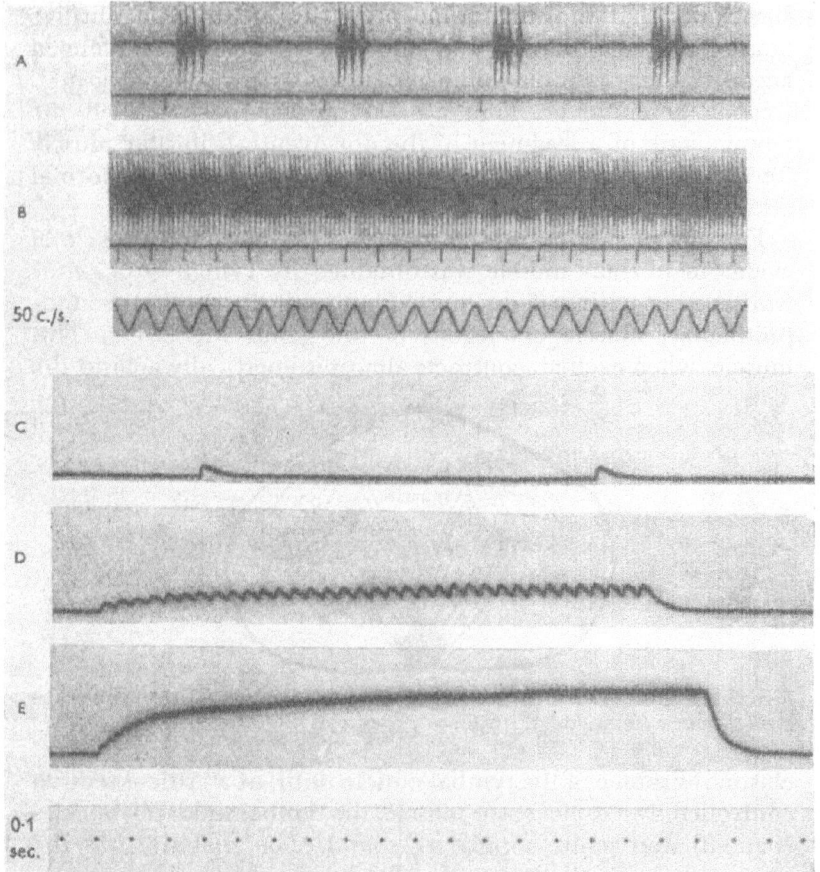

Fig. 29. Mechanical response of the tymbal muscle of *Platypleura capitata* (Cicadidae). A, B, sound pulses indicating muscle contractions in the intact preparation with quick release, stimulation at 9·5 per sec. and 47 per sec. C, D, E, near-isometric twitches and tetanus of the isolated muscle without quick release, stimulation with single stimuli, at 22 per sec. and at 46 per sec. (From Pringle, 1954*b*.)

rhythm was a deactivation of the myofibrils produced by the sudden release of tension at the IN click; this deactivation changed the state of the contractile system from that normal for an excited muscle into one in which it behaved as if relaxed and could be re-extended by the natural elasticity of the tymbal, which thus clicked back to its OUT position and restored the initial conditions. On this hypothesis, which is diagrammatized

56

in fig. 30 A, B, a distinction is necessary between the 'active state' of the fibre as a whole which results from the excitation and persists for about 40 msec. after each motor-nerve impulse, and the 'activation' of the myofibrils (indicated by tension) which becomes rhythmic if deactivation by release is present (if the

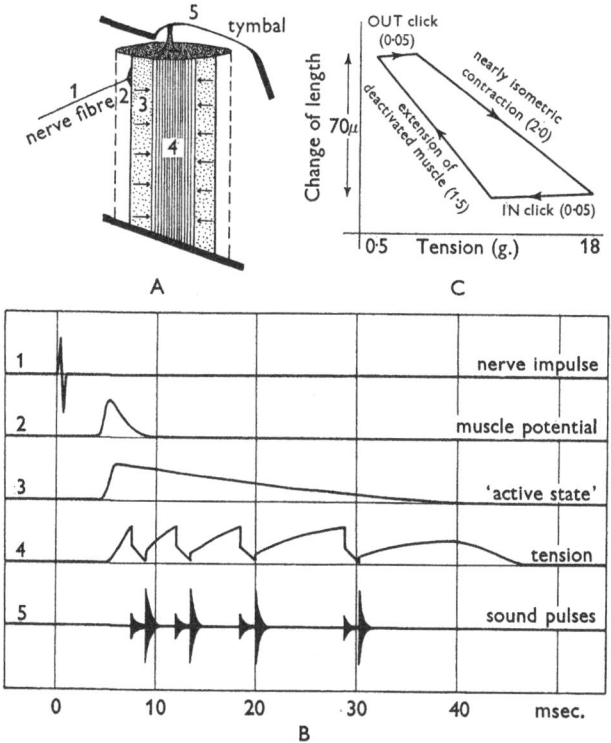

Fig. 30. A, diagrammatic representation of the sequence of events in a single tymbal muscle fibre excited from its motor nerve. B, time course of the sequence of events occurring at places numbered in A; lines 3 and 4 are interpolated according to the analysis given by Pringle (1954a). C, the energy cycle of cicada tymbal muscle contracting at 280 cycles per sec. when excited by repetitive stimulation of its motor nerve; the figures indicate the duration of each phase of the cycle in milliseconds. (From Pringle, 1956.)

muscle is attached normally to the tymbal). When motor-nerve impulses succeed one another sufficiently rapidly to maintain the active state, work is here done through a cycle of activation and deactivation (fig. 30 c) and not by repeated excitation and recovery as in normal striated muscle. This is essentially the same suggestion as that put forward by Boettiger (1951), except that the maintained condition of the muscle is identified as a persistent 'active state' and not as a tetanus, which term is commonly understood to mean a maintained tension by summation of mechanical phenomena.

In the particular case of the fibrillar tymbal muscle of *Platypleura*, deactivation by the release of tension at the IN click of the tymbal provides an adequate explanation for the maintenance of the myogenic rhythm of activity. Since the quick release was provided in Pringle's (1954a) experiments by the natural action of the tymbal, no evidence was obtained as to whether the quick stretch at the OUT click played any part in the activation cycle, or about the possible variation in energy output. By the action of the tensor muscle more or less elastic energy can be stored during the isometric phase of muscle contraction and released at the click, but, since the click action is inherent in the structure of the tymbal, there is a lower limit of energy output below which the cycle will not operate. The movement of the wings of an insect in flight is much more nearly harmonic than the movement of the tymbal, and, as was explained in Chapter 2, the extent of the departure from harmonic motion (the amount of click action) is fully under the control of certain accessory indirect muscles—the pleurosternals of Diptera and the tergopleurals of Coleoptera. This poses the question whether the quick release is essential for myogenic rhythmic activity in fibrillar flight muscle, and whether the mechanical deactivation phenomenon is a reversible one—activation by stretch as well as deactivation by release. Some evidence on these points has been obtained by Boettiger and Furshpan (1952, 1954) and Boettiger (1955, 1957a, b).

Boettiger and Furshpan (1952) used records of scutellar movement in a fly to indicate the time course of length changes in the indirect muscles, and found no quick releases of comparable duration to those which must occur in the tymbal (less than 50 μsec.; Pringle, 1954a); the oscillation might stop suddenly in less than one cycle but sometimes died away gradually over several cycles (also Roeder, 1951). In the later papers the preparation used has been the semi-isolated dorsal longitudinal muscle of *Bombus* (Hymenoptera), connected not to the natural wing mechanism but to an apparatus recording length and tension. This muscle shows the expected smooth tetanus at 30–40 stimuli per sec. when recording is isometric, but with isotonic recording may produce an essentially sinusoidal oscillation in the complete absence of a click mechanism. The observations merit some discussion, since they may make it necessary to modify the hypothesis put forward by Pringle (1954a) in its application to flight muscle.

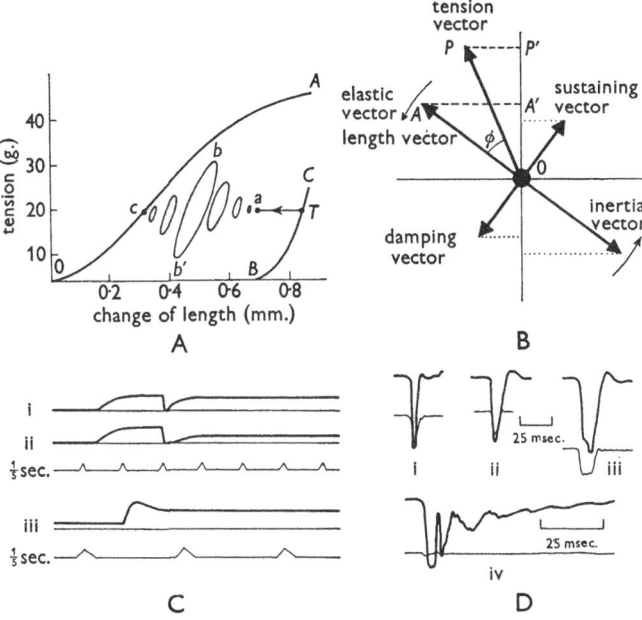

Fig. 31. Experiments with *Bombus* fibrillar muscle. A, direct cathode-ray tube plot of length/tension diagrams; *BC*, passive stretch; *OA*, isometric; *bb'*, oscillatory behaviour; for further explanation, see text. B, vector diagram to illustrate the phase relationships in a sinusoidal oscillation. C, tension changes in quick-release and quick-stretch experiments with normal (frog sartorius) muscle; i, ii, quick releases of, respectively, 4·7 and 9·7 mm.; iii, quick stretch. D, quick-release and quick-stretch experiments with *Bombus* muscle; the thin line indicates length changes and the thick line tension changes; i, control with passively stretched muscle; ii, iii, iv, excited muscle, various intervals between release and restretch. Diagrams i, ii, iii are to the same time-scale. (C redrawn from Gasser and Hill, 1924; A, B, D redrawn from Boettiger, 1957b.)

Boettiger (1957a, b) displays the experimental results as shown in fig. 31 A. The *Bombus* dorsal longitudinal indirect muscle, with its normal anterior attachment to the isolated thorax, is connected through the detached postphragma and a short length of chain to a mechano-electrical transducer measuring tension; the platform bearing the preparation is pivoted to allow it to move vertically and its displacement is measured by means of another transducer which thus gives an indication of muscle length. When the amplified outputs from the two transducers are connected respectively to the Y and X plates of a cathode-ray tube, a direct plot is obtained of the length/tension diagram of the muscle. The area between the passive stretch curve (BC) and the curve of isometric tension at various lengths (OA) represents the possible range of mechanical states of the muscle (compare fig. 26 for 'normal' muscle). When loaded to tension T with a weight and then stimulated at a sufficiently high frequency, the muscle shortens in a normal isotonic manner to point a and then oscillates with increasing amplitude to produce the steady-state curve bb'. The frequency of oscillation is determined by the inertia of the load and the elasticity of the muscle, and the counter-clockwise rotation of the spot shows that work is being done by the muscle on the external damping. If the damping is increased slightly the loop widens, showing that the energy output increases; if it is increased much, the oscillation ultimately stops and full shortening then proceeds to point c; but, on removal of the damping, oscillations recommence and grow to the same steady-state curve bb'. By adjustment of loading and stimulation, oscillations can be produced at different positions within the area of possible states.

In some of Boettiger's earlier experiments with intact flies, cases occurred in which the wing movements stopped abruptly and remained stopped for the duration of several cycles, and when oscillation restarted after removal of the inhibition to movement, the shorter of the pair of indirect muscles could be as rapidly lengthened by its antagonist as in the normal stroke. This observation and the sinusoidal form of the oscillations in isolated *Bombus* muscle convinced Boettiger that a transient deactivation by release could not be the complete explanation of the peculiar properties of fibrillar muscle, and that there must also be an unusual relationship between lengthening and the return of tension.

In any undamped elastic system tension is instantaneously related to length in a cycle of length changes. In excited fibrillar muscle Boettiger concludes that changes of tension in the contractile element are delayed after changes in length, so that the tension is higher in a muscle which is shortening than in one which is lengthening. He summarizes the forces present in the experiment with oscillating *Bombus* muscle by means of a vector diagram (fig. 31 B), based on the fundamental equation of oscillatory harmonic motion

$$m\frac{d^2x}{dt^2} + (c-s)\frac{dx}{dt} + kx = 0,$$

where x is the change in length, m is the inertia, k is the co-efficient of elasticity, and $c(dx/dt)$ and $s(dx/dt)$ are, respectively, the damping force and the sustaining force, both proportional to the velocity. The instantaneous values of each of these forces and of the muscle length are given in fig. 31 B by the projection of the vectors on the vertical axis. Since both the elastic and the sustaining forces are generated in the muscle machine the projection OP' of their resultant OP gives the total muscle tension. In the cycle produced by anti-clockwise rotation of the vectors, tension changes (represented by OP') lag behind length changes (represented by the projection OA' of the length vector OA) by an amount depending on the phase angle ϕ which is considered to change from a negative to a positive value on excitation. This representation helps one to visualize the essential feature which must be present in a self-maintaining oscillatory system doing work upon an external damping load.

Further evidence of the reality of the delay between tension changes and length changes is provided by experiments in which *Bombus* muscle is given transient quick releases and quick stretches; these may be compared with the classical experiments of Gasser and Hill (1924) on a normal striated muscle, the sartorius of the frog. If a normal muscle is stimulated to develop its isometric tetanic tension at any particular length and is then suddenly allowed to shorten through a small distance, the tension falls (to zero with sufficient release) and is then re-developed to the amount characteristic of the new length (i, ii, fig. 31 C). This behaviour is interpreted as due to the interaction of a contractile element which cannot shorten faster than a certain rate and a series elastic element which goes slack on release. If a similarly active muscle is given a small stretch at a rate not fast enough to produce irreversible damage due to 'slip' in the structure, tension rises initially above that characteristic of the new length and then declines to this value (iii, fig. 31 C). In a cycle of imposed length changes, tension is lower during shortening than during lengthening and work is always done by the external source on the muscle.

In control experiments with stretched but unexcited *Bombus* muscle (i, fig. 31D), Boettiger found that length changes and tension changes were simultaneous apart from the expected small tension overshoot after restretch. In the excited muscle given a rapid restretch (ii, fig. 31D), tension continues to rise after the original length is re-established; if restretch is delayed (iii, fig. 31D), tension is also seen to continue to fall after the end of the quick release. With an even shorter interval between release and restretch (iv, fig. 31D), the sudden (elastic ?) rise of tension during restretch is followed by a delayed fall and an even more delayed rise again followed by oscillation. While it is clear that both here and in the oscillation experiments the apparatus is not as yet technically good enough to justify truly quantitative measurements of these rapid events, the results show a property of excited fibrillar muscle quite distinct from those considered normal for striated muscle.

In spite of differences in terminology, essentially the same concept is inherent in the hypotheses of Boettiger (1957a, b) and Pringle (1954a), since a direct effect of lengthening on the activation of the contractile element was not excluded by the latter author. The almost complete deactivation by release demonstrated for the cicada tymbal muscle now appears to be a special extreme case of a property which plays a useful role in flight muscle even when the deactivation is not elicited to such a large extent, as in the sinusoidal oscillations. It remains to be shown that near-sinusoidal changes of tension are, in fact, present in the indirect muscles during flight. Boettiger (1957b), using Hocking's (1953) figure of 66 % for the aerodynamic efficiency of insect wings and data from his own experiments with isolated *Bombus* muscle, calculates that about 50,000 ergs per sec. of useful power might be generated by a sinusoidal work cycle, a figure 'somewhat low considering the large size of the bumble-bee'. He also states that experiments with a mirror on the scutellum of *Sarcophaga* (the motion of which displays directly the movements of the indirect muscles; see p. 13) show nearly sinusoidal length changes during tethered flight in this dipteran. Nevertheless, it is clear that the click mechanism in the wing articulation introduces a very non-linear element into the mechanics of motion in the intact flying insect, and it cannot therefore be assumed that tension changes in the indirect muscles will also be sinusoidal. A greater energy output could be obtained by a non-sinusoidal cycle within the available area of fig. 31 A than by any possible elliptical loop, and the isometric tension might even be momentarily exceeded during quick stretch. Control of the form of the click action by the accessory indirect (pleurosternal) muscles could therefore provide a considerable range of control of power output, and it is not to be expected that oscillatory experiments on the isolated muscles would achieve the full power available in flight.

The underlying biophysical and biochemical mechanisms responsible for the peculiar properties of fibrillar muscle are far from clear. Pringle (1954 a) pointed out that there was then no evidence that deactivation by release might not be a latent property of all striated muscle, which was realized only in these insect systems because of the presence there of suitable skeletal mechanisms. Ritchie and Wilkie (1955) have now re-examined frog sartorius muscle in the light of this suggestion but have found no sign of the effect; it remains true, however, that the oscillatory behaviour found by Gasser and Hill (1924, their fig. 9) after very rapid stretch is not satisfactorily explained in terms of an irreversible, damaging 'slip' in the muscle structure, as those authors suggest. There is recent evidence (Goodall, 1956; Lorand and Moos, 1956) that maintained auto-oscillations can occur in extracted mammalian muscle-fibre systems under conditions which preclude the involvement of rhythmic excitation. These oscillations occurred at about the natural period of the suspensory mechanical system (2·5–3·0 per sec.), and seem to that extent to be similar to the myogenic rhythms of fibrillar muscle. An inertial load appeared to be essential for the production of oscillations, but otherwise the necessary conditions were not determined. A full discussion of the implications of the myogenic rhythm on theories of the mechanism of muscular action would, however, be out of place here, and is probably undesirable until more accurate quantitative information is available about the phenomenon.

THE EVOLUTION OF FIBRILLAR MUSCLE. Roeder (1951) was the first to point out that, in the physiology of their flight muscles, insects fall into two classes—those in which there is a 1:1 relationship between muscular contractions and motor-nerve impulses, and those in which this ratio is greater than one. Boettiger (1957 *a, b*) has now established that the latter category includes only the Diptera, Hymenoptera, Coleoptera and Hemiptera other than the Cicadidae (fig. 27); apart from flight and haltere muscles, a myogenic rhythm of activity has been found only in the tymbal muscle of certain cicadas (Pringle, 1954*a*) and, probably, some jassids and cercopids (Ossiannils-son, 1949; Pringle, 1957). The correlation with histological evidence for the distribution of fibrillar muscle is good (table 1). Since there is no suggestion that these different orders of insects are derived from a common stock, it is clear that this histological and physiological specialization has occurred several times in evolutionary history, and it is instructive to consider briefly the possible stages in the process.

Of the two physiological classes, the 1 : 1 muscles are evidently primitive; histologically they differ less from normal trunk muscles and they occur in the more primitive orders. It has to be established, first, that there is a biological advantage in the myogenic system, and, secondly, that the condition can be achieved without loss of functional continuity. If, as seems probable, the wings of the first Pterygota were large tergal expansions used as fixed planes for gliding flight, their move-ments in the immediately subsequent flapping stages were slow and no problems arise about the speed of action of the muscles which moved them. With reduction in size of the wings and of the insect as a whole, this difficulty would begin to be acute when the frequency of beat needed to produce a sufficient aero-dynamic lift approached 50–70 per sec., for 15–20 msec. appears to be the minimum time required for a complete operation of a muscle working on the ordinary excitation-recovery cycle. Pringle (1957) has suggested that it is probably at this stage that the click mechanism in the wing articulation began to be developed, and that deactivation by release may play a role in assisting relaxation even in muscles of the 1 : 1 type which do not have a fully developed myogenic rhythmic mechanism. A click action occurs in at least the hind wings of *Schistocerca*, which falls into this class (Weis-Fogh, unpublished), but it has not been determined whether the release of tension which the click produces has a deactivating effect in *Schistocerca* muscle; experiments on the twitch properties of the dorsal longitudinal muscle give an indication of such an effect (Buchthal *et al.* 1957). The highest wing-beat frequencies in an order known to have flight muscles of the 1:1 type occur in the Lepidoptera; Sotavalta (1947) reports about 100 per sec. in some Sesiidae (Aegeridae), which is difficult to explain in terms of the known properties of the normal type of striated muscle.

If deactivation by release can assist relaxation in a 1:1 flight muscle system, then a mechanism could exist which is intermediate between the primitive type and the fully myogenic fibrillar system. Tiegs (1955) suggests that the pseudo-fibrillar muscle of some Homoptera represents such an intermediate stage; his argument is based on the observation that these muscles, like the tubular and close-packed types, are capable of tetanic shortening to an obvious extent under repetitive stimulation. Since all high-frequency fibrillar muscle functions under nearly isometric conditions and can shorten only to a very small extent (12 % maximum in *Bombus* dorsal longitudinal muscle, Boettiger, 1955; probably less in the higher Diptera) the point is valid, and intermediate types might be sought in species from myogenic orders whose muscles are capable of a considerable degree of shortening. For example, the Tipulidae are known to have fibrillar muscle (Tiegs, 1955) and the considerable acceleration of beat frequency produced by shortening the wings (von Buddenbrock, 1919) suggests that the rhythm is myogenic; nevertheless, electrical stimulation of the longitudinal muscle in a bisected thorax causes an easily visible contraction (Tiegs, 1955). The frequency of wing beat in this family is sufficiently low for synchronous motor-nerve impulses to occur in the normal intact insect.

Once the property of deactivation by release is established, the evolution of a truly myogenic rhythm requires only further specialization of the articular mechanism, and it then becomes advantageous for the active state following each excitation to be prolonged, so that modulation of the wing-beat frequency at the frequency of motor-nerve impulses is avoided. The indirect muscles of the higher Diptera and Hymenoptera are slow muscles, as judged by the normal activity cycle. When deactivation by release is prevented by detachment of the muscles of a fly from the skeleton, a stimulus frequency of 10–20 per sec. is adequate to produce a smooth tetanus (Boettiger, 1951); in *Calliphora* in 'anaesthetic flight' a motor-nerve impulse frequency of about 3 per sec. accompanies an exactly regular wing-beat frequency of 120 per sec. (Pringle, 1949). In this respect the Hymenoptera appear to be less advanced than the higher Diptera (data for *Vespa* in Roeder, 1951).

The number of separate evolutionary lines which have led to a fibrillar myogenic system is remarkable, and it seems that the potentiality for this behaviour must be latent in insect muscles in general, if not in all muscle. Not only in the indirect and direct thoracic muscles, but also in vertical (tymbal) and longitudinal muscles of the first abdominal segment (Delphacidae, p. 39) has the modification occurred in response to a functional need. The elucidation of the physiological and biochemical mechanisms responsible for the behaviour remains a major problem for insect physiologists.

CHAPTER 4

Aerodynamics

A heavier-than-air flying machine maintains itself in the air and moves by virtue of the differences in pressure produced by the flow of air over its surfaces. Even with objects of regular shape this air flow is complicated, and if the shape is irregular it becomes almost impossible to give a complete description of the pressure distribution under all important conditions. Direct measurements of pressure are possible only on specially constructed models, and a variety of standard procedures have therefore been developed by aeronautical engineers which enable the total forces on a structure to be determined and the results expressed in a form which is useful for theoretical and practical studies. It is convenient to begin this chapter with a brief description of these techniques and of their advantages and limitations for the student of insect flight.

PRINCIPLES. When a solid object starts to move through the air, or when the air starts to move over a stationary object, the air flow is initially different from that occurring once circulation is established. Classical aerodynamic theory deals only with the steady-state conditions. It has been argued (Osborne, 1951) that the rapid alternation of movement in the flapping flight of insects makes steady-state aerodynamics inapplicable, but the recent studies of Weis-Fogh and Jensen (1956) do not support this conclusion, at any rate so far as concerns the locust. There are indications that exceptionally large lifting forces can be generated by aerofoils during the short periods when the steady-state air flow is being established (Farren, 1936), but, since no valid theoretical framework exists for the treatment of data on rapidly accelerating wings, the zoologist may well neglect the problem unless and until it is demonstrated that such an approach is necessary for an understanding of insect flight.

In flying animals as in flying machines it is convenient to distinguish between the analysis of straight-and-level flight and problems of stability and control. In the first analysis the

67

structure is assumed to be laterally symmetrical and no account is taken of sideways drift or of rotations about the centre of gravity. If the shape is complicated it is first divided arbitrarily into elements and the aerodynamic force acting on each element is then resolved into the two components of lift and drag (fig. 32), which are found to be fully describable in terms of two coefficients C_L and C_D, defined by the equations

$$\text{Lift} = C_L . S . \tfrac{1}{2}\rho V^2,$$

$$\text{Drag} = C_D . S . \tfrac{1}{2}\rho V^2,$$

where S is the area of the element, ρ is the mass density of the air and V is the velocity of the relative wind. The coefficients of

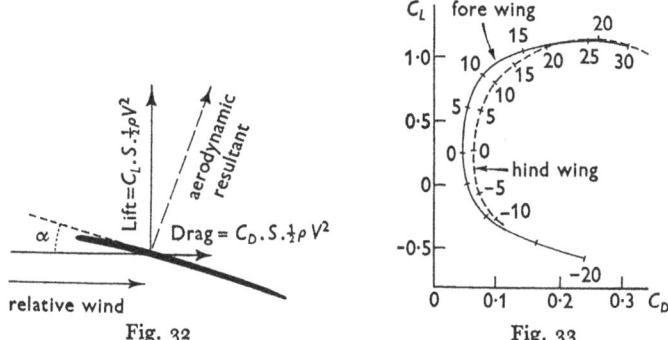

Fig. 32   Fig. 33

Fig. 32. Diagram to illustrate the resolution into lift (L) and drag (D) of the aerodynamic resultant force acting on a wing element at angle of attack α to the relative wind.

Fig. 33. Polar curves showing the measured coefficients of lift and drag of *Schistocerca* wings in a graded wind field simulating angular motion. The angle of attack is shown by figures on the curves. (Redrawn from Jensen, 1956.)

lift and drag change according to angle of attack α between the relative wind and the plane in which the area is measured, and a graphical plot of this relationship (fig. 33) gives the best concise statement of the aerodynamic properties of the structural element. The properties of the whole structure can then be derived from those of its constituent elements with due allowance for their mutual interaction.

In flapping flight, which is the important case for the great majority of insects, the main aerodynamic surfaces, the wings, move in relation to the main load, the body. The average aerodynamic force on the wings may therefore act in such a direction as to have a positive forward as well as a positive

68

upward component, so that a thrust as well as lift is generated at the wing base. The lift and drag totals of the whole insect are the vector sums of the lift and drag of the non-moving body and of the moving wings; the sum of all forces is zero in an insect in straight-and-level flight at constant velocity. Measurements can be made directly of the total aerodynamic forces if the insect is fixed to a balance and suspended in an air stream, or, alternatively, the aerodynamic properties of the body and wings can be measured separately and the total forces calculated from a knowledge of the kinematics. In the latter case allowance has to be made for the fact that the motion of the wings is angular, with a gradient of velocity from base to tip, and also for the fact that the relative wind is the vector sum of the wind due to the translational velocity of the insect as a whole, that due to the flapping and the induced wind produced as the average result of the wing motion.

Scale affects aerodynamic performance, the form of the air flow depending on the value of the Reynolds number, defined as

$$Re = \frac{\rho V d}{\mu},$$

where d is a measure of linear dimensions and μ is the viscosity. The coefficients C_L and C_D change only slowly as Re is reduced from a high value to about 1000, but at very low values C_D increases rapidly (Thom and Swart, 1940). This scale effect has to be taken into account in comparing the performance of insects and man-made machines; Re for aeroplanes is usually greater than 10^6, whereas for *Schistocerca* it is about 2000 (Jensen, 1956). It may introduce entirely new concepts for the flight of the smallest insects (Horridge, 1956; see p. 77).

HISTORY. A full review of experimental work and theories about the aerodynamics of insect flight has been given by Weis-Fogh and Jensen (1956). These authors point out the inadequacy of many attempts to understand the nature and magnitude of the aerodynamic forces which have been based on incomplete data and erroneous concepts about the principles of the subject. Unfortunately, large errors are liable to be introduced by over-simplification; thus Chadwick (1953), using the formula

$$R \propto (V_{rw})^2 \sin \alpha$$

for estimating the aerodynamic resultant force R in terms of the relative wind V_{rw} and the angle of attack α, finds that, when he substitutes figures for *Drosophila* based on his own observations and those of Magnan (1934), the vertical lift, integrated over the whole cycle, is insufficient to balance the weight of the insect. He quotes Magnan for the suggestion that the deficit is made up by the 'region of greatly reduced pressure' between the wing surfaces when they are closely approximated above the body at the start of the downstroke. In fact, the value of $(C_L^2 + C_D^2)^{\frac{1}{2}}$, which should be used in the formula in place of sin α, is probably about five times larger for thin-cambered aerofoils at the angles of attack concerned, and there are no grounds in Magnan's data for assuming that non-steady-state conditions at the start of the downstroke make a significant contribution to the aerodynamic force.

It was pointed out in Chapter 2 that the form of the wing beat of a tethered insect is different from that adopted during free hovering or free forward flight; there is good experimental evidence (p. 110) that the angle of attack of the wings at different parts of the stroke is under reflex control by means of the direct muscles, and other parameters can probably also be varied in relation to the sensory inflow. One cannot therefore safely use measurements made on tethered insects in still air for quantitative evaluation of the aerodynamic forces or of the 'aerodynamic efficiency' of flight.

Conceptual or methodological imperfections of this sort are regarded by Weis-Fogh and Jensen (1956) as sufficient to render of little value the theoretical conclusions of Demoll (1918, 1919), Magnan and Sainte-Laguë (1933), Attila (1947), Sotavalta (1952) and Hocking (1953). On the other hand, the theories about flapping flight of Holst and Küchemann (1941, 1942), Walker (1925, 1927) and Osborne (1951) are capable of giving valuable insight into the factors involved, though each suffers from certain serious limitations. In order to test them to the full, Weis-Fogh and Jensen have invented the three idealized insect types shown in fig. 34, and have inserted into the theoretical formulae values for the important parameters derived from their own observations ('locust type') and those of Hocking (1953) ('horsefly' and 'mosquito' types); in both these experimental studies, measurements were made on insects flying under conditions comparable to those of free flight, Weis-Fogh's in a wind tunnel and Hocking's on a roundabout. The conclusions may be summarized as follows.

In Holst and Küchemann's (1941) theory the pair of wings is treated as a single aerofoil of double span moving with a constant translational velocity upon which is superimposed a vertical sinusoidal oscillation of amplitude a; the angle of attack also oscillates sinusoidally about a mean value α_o with an amplitude α_a and a phase difference Φ with the vertical oscillation of the wing. The angular motion of the wings is thus replaced by an overall displacement. If certain reasonable assumptions are made about the relationship between C_L and the angle of attack, it is then possible to derive formulae for the mean lift and thrust. The formulae are manageable only if the

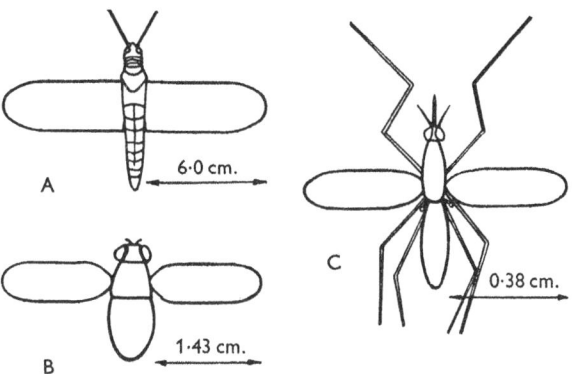

Fig. 34. The idealized types used by Weis-Fogh and Jensen (1956) to evaluate the merit of theories about insect flight. A, 'locust' type; B, 'horse-fly' type; C, 'mosquito' type.

oscillatory velocities are small compared to the translational velocity, as in many birds, and quite unrealistic results are obtained by inserting values for the 'horse-fly' and 'mosquito' types. The most favourable conditions for generating both large lift and thrust are those where $\Phi = \frac{1}{2}\pi$ and $\alpha_a = \alpha_o$; the angle of attack is then α_o at the beginning of the downstroke, $2\alpha_o$ when the wing passes the middle position, α_o at the bottom and zero at the middle of the upstroke. This corresponds approximately to the state of affairs found in the locust (Jensen, 1956), and the insertion of parameter values for the 'locust' type leads to values for the total thrust and for the angles of attack which are reasonably close to those observed. The theory therefore has some value as a simplification for slow-flapping insects; it is quite unable to provide an explanation of fast-flapping or hovering flight.

Walker's (1925, 1927) theory was developed to explain the forward flapping flight of birds, but Weis-Fogh and Jensen (1956) have adapted it to apply to certain insects with interesting results. It simplifies the flight situation by supposing that (a) a positive translational velocity exists, (b) the angular velocity of the wings is constant during up- and downstrokes, (c) the plane of the oscillation is vertical, (d) the angle of attack is constant during one half-cycle but differs during up- and downstrokes. The induced wind (the average air motion resulting from the flapping) is ignored. In order to allow for twisting of the wings, the span is arbitrarily divided into three sections, to each of which constant angles of attack are assigned during the up- and downstrokes; in Walker's own example of the rook, reasonable values of total lift and thrust were obtained by assuming that the wing section had the characteristics of a thin aerofoil and that its angle of attack during the downstroke was 12° over the whole span, but that during the upstroke it was supinated so that the tip had zero incidence. With the values adopted for the 'locust' type, with figures for C_L and C_D averaged between measurements on the fore and hind wings (fig. 33), and with a similar supination twist during the upstroke, Walker's theory gave a reasonable value of total thrust (92 mg. as against 85 mg. measured extra-to-wing drag in steady flight) but insufficient lift (1750 mg. as against 2000 mg. standard body weight); if, however, allowance was made for the known 30° inclination of the plane of the wing beat to the vertical in the locust, the theoretical value for total lift became sufficient to sustain the body weight. The interesting conclusion emerged that three-quarters of the lift is produced by the inner two-thirds of the wing, the middle region being the most important. The thrust, on the other hand, depended entirely on the outer third of the wing, the middle region being neutral and the inner third contributing to the drag. Provided that allowance was made for the stroke-plane angle, the theory also gave reasonable average values for lift and thrust for the 'horse-fly' type, in which the flapping speed of the wing tip is 2·75 times the forward speed ('locust' type 1·2), but in the 'mosquito' type, in which the ratio is 4·9, calculated values for total lift and thrust were insufficient, due probably to the theory's neglect of the induced wind.

Osborne's (1951) theory differs from the other two in that no assumptions are made about the changes in angle of attack during the different parts of the wing stroke. Instead, using a formula for the forces acting on a wing-surface element moved in a sinusoidal manner, the average lift and thrust are calculated on the assumption that the force is proportional to the square of the relative wind. The mathematics are complicated, and in order to relate lift and thrust to the average C_L and C_D it is necessary to introduce 'shrewdly chosen averages' in place of integrals over span and wing-beat cycle. An expression is also obtained for the minimum average aerodynamic resultant $(C_D^2 + C_L^2)^{\frac{1}{2}}$ min., this time by exact integration. When observed values of total lift and thrust are inserted in these formulae, average values of C_L and C_D and of the minimum average aerodynamic resultant are obtained for a variety of insects. Account is taken of the induced wind, which is calculated according to a momentum theory developed for helicopter rotors (Glauert, 1935) and is assumed to be uniform over the span. For reasons which are not clear this procedure gave highly aberrant values for the average lift and drag coefficients if account was taken of both wing strokes, and Osborne therefore assumed that lift and thrust were generated only on the down-stroke; he took his experimental data from Magnan (1934) and found a range of average C_L values varying from 0·1 for *Aeshna mixta* (Odonata) to 4–5 for *Lucanus cervus* (Coleoptera). Since a number of different insects gave values above 2·0, which is the maximum for thin aerofoils according to classical aerodynamic theory, Osborne concluded that steady-state conditions are not established in the flight of rapidly flapping insects. Weis-Fogh and Jensen (1956) criticize this argument mainly on the grounds that the data of Magnan (1934) for flying speed, wing-beat frequency and stroke angle were not obtained from simultaneous observations on the same insect and are therefore unreliable for a calculation of this sort. When they inserted values for their three insect types, reasonable values for average C_L emerged in spite of the fact that the downstroke was still ignored. Osborne's theory cannot by its nature lead to a better understanding of the instantaneous forces on the wings, but it may still be useful if and when accurate observational data are available for actual, as opposed to hypothetical, small insects.

In summary, it must be stated that there is at present no

simplified picture of the mechanism of insect flight which can be used to derive a mathematical theory of the way in which lift and thrust are generated by the wing motion. It is perhaps not surprising that this should be the case, since even for the much simpler dynamics of aquatic locomotion theoretical analyses are far from complete (Taylor, 1952; Gray and Hancock, 1955). In view of the many competent attempts which have been made to solve the problem, this means that the departures from a mathematically simple time sequence in the various aspects of the wing movement are probably significant for insect flight, and that there is no short cut to the laborious process of direct calculation, using accurate measurements of the kinematics and of the steady-state aerodynamics over the whole significant range. Such a programme has been carried through only once, for the locust *Schistocerca gregaria* by Weis-Fogh (1956*a*) and Jensen (1956); their methods and results are of sufficient interest to describe in some detail.

LOCUST FLIGHT. Jensen (1956) first used stroboscopic cinematograph films of locusts flying under controlled conditions in a wind tunnel to plot the projection of the fore and hind wing-tip paths on the three planes of reference, and to determine the wing sections and angles of attack at the base, middle and tip of the wing at about twenty instants during the stroke. By making a large number of such observations on different locusts and under different conditions (Weis-Fogh, 1956*a*) it was possible to select a few experiments truly representative of the normal performance. To facilitate analysis the wing-tip path was then transferred to a cylindrical projection from which, when developed into a plane, direct measurements could be made of the velocity of the tip and of the angular velocity and acceleration of the wing. Jensen next cut off the wings, set them in the correct section and inserted them into the boundary layer at the side of a specially constructed wind tunnel so that the gradient of velocity from base to tip corresponded approximately to that resulting from the angular motion. The polar plots of C_L and C_D of fig. 33 were obtained in this way, the Reynolds number being correct in the measurements. This method could not, however, reproduce the variation in angle of attack along the span which was found to occur in actual flight; to arrive at a figure for the lift and drag of the twisted wing, weighted averages had to be calculated from the C_L and C_D

74

values found for the observed angles of attack at base, middle and tip. The weighting factors were based on an expression for the distribution of load along the span which took account of the plan form and the gradient of velocity of the relative wind; this expression is accurate over the observed range of flying conditions to within 1 % for the fore wings and 6 % for the hind wings. A calculated correction was introduced for the interaction of the two pairs of wings. It was then possible to calculate the instantaneous values of the aerodynamic forces and, after subtracting the component acting along the wing axis, also the values of the torques about the wing base due to the aerodynamic forces; in fig. 35 these are shown resolved into com-

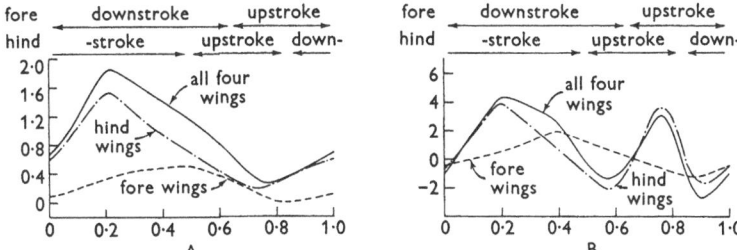

Fig. 35. Calculated aerodynamic lift (A) and thrust (B) generated by the two fore wings, two hind wings and all four wings of the locust, *Schistocerca gregaria*, in normal flight. Abscissa: time in fractions of one wing-beat cycle. Ordinate: lift or thrust in relation to the average lift or thrust of all four wings.

ponents of vertical (lift) and horizontal (thrust) force produced on the body by the fore and hind wings separately and by both pairs of wings. When summated over the whole stroke the hind wings produced 71 % of the lift; about 20 % of the total lift of both wings was produced during the upstrokes. Forward thrust was produced by the fore wings during the downstroke only, but by the hind wings during both strokes. The total average thrust was only 7 % of the average lift.

The calculated values of average lift and thrust were next compared with the measurements made in the wind-tunnel balance at the same time as the cinematograph films were taken. Four independent experiments were analysed in this way, two in which the measured lift was approximately equal to the standard value (Weis-Fogh, 1956a) and one each in which it was in excess or deficit. In both normal-lift experiments and in the high-lift case the calculated and measured values agreed to within 7% for lift and 12% for the much smaller drag; in the low-lift experiment many of the flight parameters were different from normal and the disagreement of 19% was taken to indicate that it did not represent a normal performance. The close correspondence found in the other three experiments is good evidence that steady-state aerodynamics provides a satisfactory picture of locust flight.

The changes in the angle of attack during the stroke could best be visualized from measurements made at the mid-point of the wings (fig. 4); the supination during the upstroke is well displayed. It is interesting to note that in the experiment in which exceptionally high lift was generated there was a significantly greater angle of attack in both wings throughout the beat. This supports the conclusion reached by Weis-Fogh (1956a) by elimination of all other possible parameter changes that the main factor responsible for the adjustment of lift is wing-twisting rather than any of the other variables in the flight system. Weis-Fogh found, in a series of sixty-two wind-tunnel experiments comprising 690 sets of observations, that there was little or no change in stroke amplitude, stroke-plane angle, the middle position of the wings or the time course of wing movements, and that, although the wing-beat frequency showed a correlation with lift, the magnitude of the increase in frequency was quite insufficient to account for the increased lift. The adjustment of lift is further discussed in Chapter 5 in relation to flight reflexes.

In addition to all the other parameters which change during the stroke, Jensen (1956) showed that in the locust there is also a change in the wing section. At the beginning of the downstroke the fore wing is a very slightly cambered sheet of negligible thickness except for the veins, which project on average 0·003 cm. (maximum at base 0·007 cm.) above the surface but are almost entirely within the boundary layer at the found values of the relative wind. Towards the end of the downstroke a longitudinal fold comes into operation (fig. 4), creating a flap which increases C_L at the expense of a decrease (misprinted 'increase' in the original paper) of the lift-drag ratio; since the wing has to be decelerated in this phase of the stroke, kinetic energy is usefully dissipated in this way. On the upstroke the longitudinal folding changes, producing a Z-section with moderate values of C_L and C_D over a wide range of angles of attack. All these changes of section of the fore wing are actively produced by the basal articulation; by contrast, the hind wing has a large vannal area which is passively deflected by the air stream so that a smooth camber is always retained. It is very probable that changes in wing section during the stroke occur in the majority of insects, and the results emphasize again the difficulty of arriving at a satisfactory analysis of the aerodynamics of insect flight from anything but a full investigation of the kinematics. The slight importance of surface irregularities because of the relatively thick boundary layer helps to explain the construction of wings such as those of Lepidoptera and Trichoptera in which the surface is roughened by the presence of scales or extended by dense rows of hairs.

FLIGHT OF VERY SMALL INSECTS. The special aerodynamic problems encountered by insects of very small size have been discussed by Horridge (1956). It is known from the work of Thom and Swart (1940) that when the Reynolds number is below about 200 the steady-state properties of aerofoils begin to change (fig. 36); the angle of attack for the highest lift/drag ratio increases from the usual value of about 15–20° up to a value as great as 45°, and at still lower numbers C_D increases rapidly, becoming almost independent of the angle of attack, until it exceeds by a factor of 5–10 the highest value of C_L. If, as Horridge suggests, there are insects which can fly with Re in this range, they must be doing so not by using their wings as a conventional aerofoil, but by 'rowing' themselves through the air with different C_D values on the up- and downstrokes. He calculates that a C_D ratio of 5:4 during the two strokes would produce an average upward force of the same order of magnitude as the lift from any possible aerofoil, and suggests that a possible mechanism would be to have bristles or a whole wing that bends more easily on the up- than on the downstroke. The small Hymenoptera which use their wings to swim under water (Matheson and Crosby, 1912) must generate thrust in some such way, but in the air, where lift is required all the time equal to the insect's weight, the mechanism would be extremely wasteful of energy.

While Horridge's deductions are undoubtedly valid, it does not seem to be established that there are any insects for whose flight the Reynolds number is so low that the argument is applicable. The chief difficulty in arriving at a correct figure is uncertainty about the wing-beat frequency of the examples he cites. One small insect, *Forcipomyia* sp. (Ceratopogonidae, Diptera), beats its wings at about 1000 strokes per sec. (Sotavalta, 1953), and it is possible that the Hymenoptera and Coleoptera figured by Horridge come in the same range, since they also have fibrillar wing muscles. Estimates of the Reynolds number for a ceratopogonid (Weis-Fogh, unpublished)

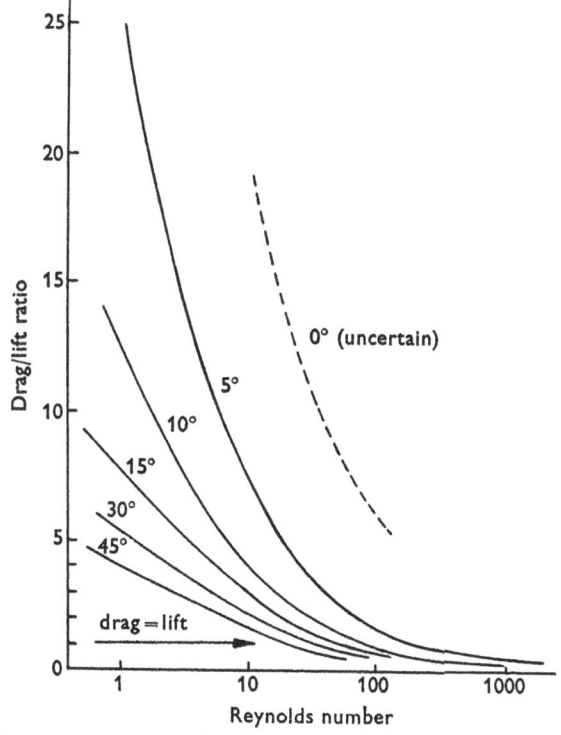

Fig. 36. The ratio of drag to lift for a thin aerofoil at various values of the Reynolds number for different angles of attack. (From Horridge, 1956.)

give a value of 150, and for *Trichogramma* (Hymenoptera) 90–100 if the wing-beat frequency is 1000 per sec.; these estimates are obtained by assuming a uniform velocity of wing movement on up- and downstrokes and an approximate value for the induced wind. Horridge appears to assume a maximum wing-beat frequency of 250 per sec., which for *Trichogramma* would give a value of 20–25 for *Re*. The difference is critical.

Whether or not the flight of small insects depends on aerodynamic lift from their wings, an explanation is needed of the regular occurrence among them of wings of the type of fig. 37. As Horridge points out, this is an evolutionary convergence which must have a useful function. His alternative suggestion seems very probable: that the location of one or more supporting veins near the leading edge of the wing allows it to assume its optimum angle of attack by the passive twisting action of the air forces working against a torsional elasticity of the wing base

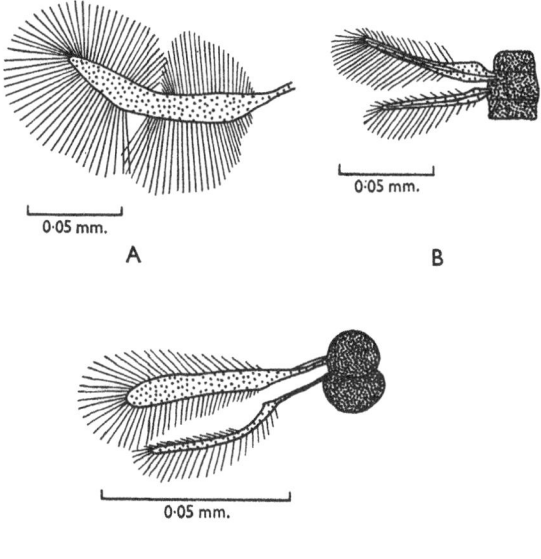

Fig. 37. The wings of some very small insects. A, *Acrotrichis* (*Trichopteryx*) sp. (Coleoptera, Ptiliidae) (redrawn from Matthews, 1872); B, *Heliothrips haemorrhoidalis* (Thysanoptera) (redrawn from Russell, 1912); C, *Patasson crassicornis* (Hymenoptera, Mymaridae) (original; *ex coll.* G. Salt).

rather than by an active twisting through the basal articulation. As has already been pointed out, the boundary layer will confer on such a wing properties similar to those of a continuous surface. Since at Reynolds numbers of about 100 the highest lift/drag ratio is achieved at high angles of attack, it might be expected that very small insects would have the supporting wing vein nearer to the middle of the chord than in those of more normal size; the centre of pressure of an aerofoil is usually one-

79

third to two-fifths of the chord back from the leading edge, and the amount of feathering action will depend on the location of the supporting vein in relation to this centre of pressure.

POWER. A considerable number of workers on the subject of insect flight have made it their main objective to determine the aerodynamic power output and, by comparing it to the oxygen consumption, to estimate the 'efficiency' of the flight system. It will be as well to consider first the nature of the problem, in order to have a clear idea of what can be achieved in an investigation of this sort. The reasoning which follows is based on that developed by Weis-Fogh and Jensen (1956).

It is possible to define 'efficiency' in several ways. In physiology the term 'muscular efficiency' means the ratio of the external mechanical work of groups of muscles to their total energy consumption. As applied to flying insects the muscular efficiency of the flight muscles is the ratio of total extra-muscular work to total energy consumption. In some investigations the attempt has been made to define an 'over-all efficiency' of flight, including both the muscular efficiency and the 'aerodynamic efficiency', which is the ratio of aerodynamic power to the total mechanical power output. Aerodynamic efficiency is a valid concept only if it refers to the total aerodynamic work per unit time; it is theoretically unsound to try to distinguish the 'useful power' (i.e. that producing forward motion; Hocking, 1953) from the 'power required for support' (also Sotavalta, 1953). The ideal flying machine requires no power for support since lift is developed at right angles to the air flow, and, since in insects lift and thrust are generated by the same effector system, determination of the one without the other has purely empirical value. Similar criticisms apply to the conclusions of Chadwick (1953) which are based on measurements of the 'useful' work done by tethered insects. If, as in the work of Hocking (1953), one of the main objectives is to determine the horizontal distance which an insect can travel in still air on a certain amount of muscular fuel, then it may be reasonable to ignore part of the aerodynamic work, but such studies cannot form the basis for an understanding of the total energy balance sheet. Mention has already been made of the dangers of drawing conclusions about energy consumption from results on fixed insects which are likely to use their wings in a manner very different from that found in free flight.

A number of investigators have determined the relationship between the wing-stroke frequency and other parameters of the flight system, and from straight-line graphs drawn through logarithmic plots of the experimental results have arrived at a figure for the exponent. Thus Sotavalta (1952), studying various insects in free flight, finds that frequency varies as (wing inertia) $^{-0.35\,(aver.)}$ and (air pressure) $^{0\,to\,0.25}$; on tethered insects Chadwick and Williams (1949) find for *Drosophila* that frequency and stroke amplitude both vary as (air density) $^{0.46}$, and Danzer (1956) finds for various Muscidae that frequency varies as (inertia) $^{-0.22}$. Sotavalta and Chadwick have each developed simplified theories about the effects which various factors should have on the power output, and have used the figures for the exponents to support conclusions about the physiological mechanisms responsible for the observed relationships on the assumption that power output is constant. It is difficult to place much reliance on arguments of this sort owing to the number of simplifying assumptions which are required; for example, both theories presuppose a sinusoidal wing motion, and neither produces convincing evidence that the power output does in fact remain constant under the conditions of the experiments.

A more promising approach to the energetic analysis of insect flight is outlined by Weis-Fogh and Jensen (1956). These authors conclude that it is unsafe to make any simplifying assumptions about the kinematics of the wing motion, and imply that the mechanical power output P must be determined by numerical evaluation of the integral

$$P = n \int_0^{1/n} Q \frac{d\gamma}{dt}\, dt,$$

where n is the wing-beat frequency, and Q and $d\gamma/dt$ are instantaneous values respectively of the torque at the wing hinge and the angular velocity of the wing. This formula is exact, but can be used only if accurate data are available for the variation with time of the total torque and the wing position. In locusts $d\gamma/dt$ can be evaluated from the curves of wing movement in the stroke plane; the torque Q, however, has to be computed from the sum of its various component parts, which must now be discussed.

At any instant in the wing stroke the sum of the torque contributions about the fulcrum must be zero:

$$Q_m + (Q_a + Q_i + Q_e + Q_d) = 0,$$

where Q_m is the torque produced by the flight muscles, Q_a is the aerodynamic torque, Q_i is the inertial torque, Q_e is the elastic torque and Q_d is the damping torque. In locusts Q_d is negligible, and the large air spaces in the thoraces of all insects with rapid wing beats make it probable that this is a general conclusion. The other three terms within the bracket are all significant in locusts (Weis-Fogh and Jensen, 1956) and are probably so in most insects. Previous workers have tended to neglect one or more terms in the equation.

The evaluation of the three main components of the torque is a matter of considerable difficulty, since it is necessary to know their variation with time unless a simple function is assumed;

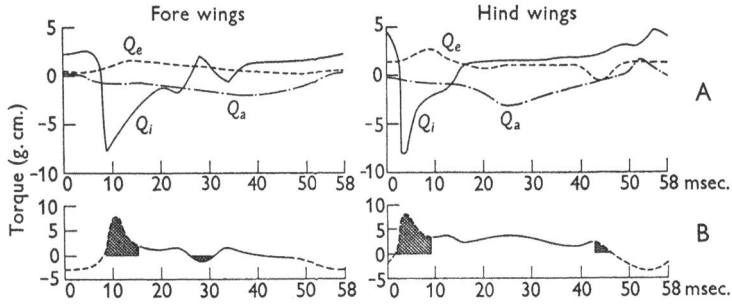

Fig. 38. A, the aerodynamic (Q_a), inertial (Q_i) and elastic (Q_e) torques at the base of the fore and hind wings of *Schistocerca* during a single stroke; the sign is negative when the torque tends to move the wing upwards. B, the total torque generated by the wing muscles; solid sections indicate downstroke, interrupted sections indicate upstroke; the hatched areas show portions of the cycle during which the muscles are doing negative work. (Redrawn from original manuscript of Weis-Fogh, 1956*d*.)

fig. 38 A shows how very far from simple the variation of Q_a is in the only experiments in which it has been determined accurately. Q_i can be calculated provided the moment of inertia and the angular motion of the wings is known. Q_e must be determined experimentally from static or slow-speed measurements with the wing in its various positions. Fig. 38 B shows the final result of approximate computations for the locust.

The other side of the energy balance sheet for flight is more easily completed since, if the nature of the fuel used by the muscles is known, the total energy consumption may be calculated from measurements of the oxygen uptake. Weis-Fogh

(1952) gives a table of the reliable measurements, ranging (in calories per gram of muscle per hour) from 400 to 800 for *Schistocerca gregaria*, 650 for *Drosophila repleta*, 1700 for *Lucilia sericata*, to 2400 for *Apis mellifera*. In *Schistocerca* the wing muscles constitute 18 % of the body weight, so that a figure of 70 cal./g. body weight/hr. may be taken as the normal rate of energy production of a steadily flying locust.

Unfortunately, it is still not possible to arrive at a reliable figure for the muscular efficiency of the flight muscles, for, as fig. 38 B shows, during part of the wing-beat cycle the muscles are doing negative work, i.e. they are actively slowing down the wing motion at the end of the up- or downstrokes. There is good evidence that in vertebrate muscle the energy supply required for negative work is considerably less than that required for positive work (Abbot, Bigland and Ritchie, 1952). Weis-Fogh (1956 *d*) estimates that of the total mechanical power generated by the locust (13·7 cal./g. body weight/hr.), 8·9 cal. go into positive work and 4·8 cal. into negative work. If the proportional efficiency of locust muscle is the same as that of human muscle, the muscular efficiency in the locust is 14 %; if negative work is as expensive as positive work, then the figure is 20 %. These figures are comparable to those obtained for vertebrate muscle working under optimum conditions, and are much larger than those computed by Chadwick (1953) for *Drosophila* and *Vanessa*; it has already been pointed out, however, that the low estimates of 3–4 % obtained by Chadwick are based on inadequate data and dubious theoretical arguments.

The demonstration that elastic and inertial torques play a significant role in the energetics of locust flight implies that there is a considerable storage of energy in the mechanical system, and contrasts with the conclusion of Sotavalta (1952) that the work done by the muscles is entirely dissipated at each beat. Sotavalta's argument is inconsistent in that he bases his reasoning on the existence of sinusoidal motion which is to be expected only in an undamped resonant system; his evidence, which is derived solely from the observed relationship between wing inertia and wing-beat frequency, is inadequate to support the conclusion. Unfortunately, there is no extensive information about the kinematics of the wing beat in insects other than the locust and about the way the wing motion changes under different conditions; in Diptera, as mentioned on p. 10, the evidence suggests that the motion is somewhere between sinusoidal and linear during each stroke. If it is generally true that all three terms, Q_a, Q_i and Q_e, are significant in the expression for total torque, then the motion should be that of a highly damped resonant system, and neither the sinusoidal nor the linear simplification will be fully adequate as a basis for theoretical arguments.

A somewhat similar conclusion may be drawn from the experiments of Danzer (1956), in which the thorax, wings and halteres of various insects were studied as a mechanically resonant system. Danzer used a powerful electromagnetic transducer to impose forced vibrations on the whole insect under various conditions and found that the frequency of wing beats in a semi-anaesthetized fly could be changed to that of the imposed vibration when this was within about 40 cycles per sec. of the normal frequency. Some motion of the wings and almost full oscillation of the halteres could also be produced artificially in a fully anaesthetized fly in which the muscular driving mechanism was not in action, but the resonance of the wings was not a sharp one and depended on wing position; the halteres showed a narrow resonance peak, but the change of resonant frequency with amplitude showed that non-linear elements were involved. With the wings fixed the thorax itself had no resonance in the significant range. Since there is good evidence that in a fly beating its wings in 'anaesthetic flight' ('Rauschflug') reflexes are inoperative, Danzer concluded that the thorax-wing system with active flight muscles does comprise a resonant system, in which the muscles behave as a 'Kippsystem' whose resonant frequency depends on muscle length, loading and metabolism. The meaning of this term is not very clear, but it seems to be implied that the muscles deliver energy discontinuously whenever some critical level is reached in the cycle of oscillation; the concept is not in conflict with the views about the physiology of fibrillar muscle put forward in Chapter 3.

CONCLUSION. Although we have only the beginnings of an understanding of the aerodynamics and energetics of insect flight, it is clear that the stage has now passed when it is of any value to put forward hypotheses based on a simplification of the flight system. For all but perhaps the smallest insects and those with the highest rates of wing beat, it seems likely that classical aerodynamic theory can be used in a quantitative analysis, and in order to demonstrate that wing accelerations make a significant contribution to the lift in the remainder it will be necessary first to prove the inadequacy of normal methods of treatment. Further observations on very small insects are needed to determine whether unusual properties are to be expected from their wing action. As with many other branches of comparative physiology, it may prove to be more valuable to concentrate effort on obtaining a very complete picture of the flight system in a small number of selected types, rather than to attempt any comprehensive survey with the inevitable danger that generic differences will preclude a general synthesis. The most urgent need is for a detailed study of the flight of one of the larger Diptera or Hymenoptera on the lines adopted by Weis-Fogh (1956a) and Jensen (1956) for the locust. Purely anatomical and behavioural studies might then suffice to extend the reasoning over a wide range of types.

Evaluation of the energy balance sheet is part of the analysis and forms an integral part of studies of the aerodynamics on the one hand, and of the physiology of the flight muscles on the other. Too much reliance should not be placed on estimates of 'efficiency' in the flight system until more is known about these different aspects in several chosen types.

CHAPTER 5

Nervous and Sensory Mechanisms

WITH few exceptions the skeletal and muscular components of the insect flight system are restricted to the second and third thoracic segments. Flight is, however, influenced and to some extent co-ordinated by stimuli which affect other parts of the insect's body, and in this chapter it is necessary to make a wider survey than has so far been required.

ANATOMY. The synaptic regions of the nervous system of a primitive insect are largely restricted to the chain of ventral ganglia, each of which has direct control of much of the activity of the segment in which it lies; even when the ganglia are fused into a smaller number of morphologically distinct structures it is often possible in histological section to distinguish their component regions. The pattern of nerves leaving the ganglia in the thoracic segments varies considerably, and details of innervation have been worked out in only a few examples. Fig. 39A shows the general plan of the nervous system in the metathorax of *Locusta*; in both meso- and metathorax fibres join the anterior wing nerve (N1) from the ganglion of the segment in front (Albrecht, 1953). In the metathorax of *Acanthacris ruficornis* (Acrididae) the indirect dorsal longitudinal and tergosternal muscles are supplied by the most anterior nerves from the ganglion (which is joined by a recurrent nerve from the mesothorax), but branches to some of the direct wing muscles come from the more posterior roots (Ewer, 1953, 1954*a*). By way of contrast, fig. 39B shows the thoracic nervous system of *Drosophila*, which is very similar to that of *Calliphora* (Lowne, 1890). Here the three pairs of nerves to the legs (I_v, II_v, III_v) leave the ganglion ventrolaterally, and the wings, halteres and flight muscles are innervated by large dorsal and smaller accessory nerves from the middle and posterior regions. Lowne mentions that the large dorsal wing nerve (II_d) enters the neuropile by two roots, of which only the ventral is represented in the origins of the haltere nerve (III_d); since the latter is purely

86

sensory (Pringle, 1948) this conforms to the generally accepted plan of organization of insect ganglia (Wigglesworth, 1950) in which the motor neuropile is located dorsally. A similar reduction of the dorsal motor root is seen in the mesothorax of beetles (Binet, 1894).

MOTOR NERVE FIBRES. As opposed to the gross anatomy of the nerve trunks which can be revealed by dissection, we have knowledge of the pattern of distribution of motor nerve fibres to

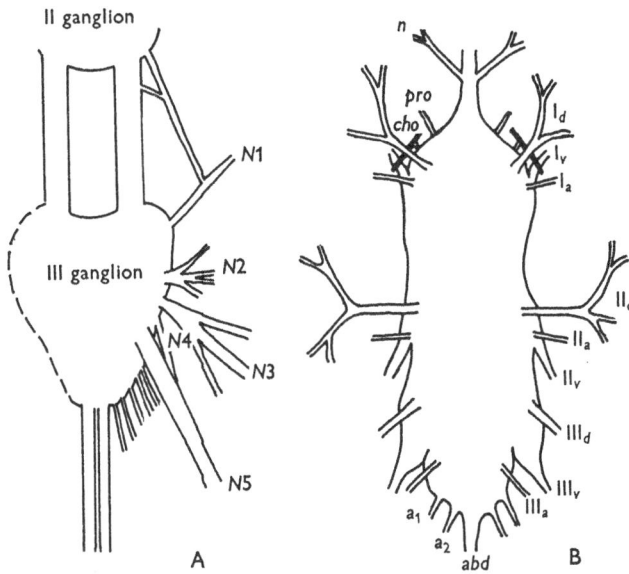

Fig. 39. A, metathoracic nerve trunks of *Locusta migratoria* (diagrammatic dorsal view); $N1$, $N2$, nerves to wing and flight muscles; $N3$, $N4$, $N5$, nerves to leg and leg muscles (redrawn from Hoyle, 1956a). B, dorsal view of thoracic ganglion of *Drosophila melanogaster*; a_1, a_2, nerves to first and second abdominal segments; *abd*, abdominal nerve cord; *cho*, chordotonal organ; *n*, neck nerve; *pro*, nerve to prosternal organ; I, II, III, nerves of pro-, meso- and metathorax, I_a, accessory, I_d, dorsal, I_v, ventral. (Redrawn from Hertweck, 1931.)

the wing muscles in only one insect (*Telea polyphemus*, Lepidoptera). Nuesch (1954) studied this question by cutting various nerve trunks in the pupa, and found that muscles innervated by fibres in these trunks then failed to develop in the imago. His results are shown in fig. 40. Nerve fibres originating in the prothoracic ganglion (I) innervate the oblique dorsal muscles (dl_2, dl_3) and the lower part of the dorsal longitudinal muscles (dl_1) in the mesothorax; the upper part of the dorsal longitudinal muscle and a small intersegmental muscle (I*is*) are innervated through the anterior root of the mesothoracic ganglion. Certain pleural muscles receive motor fibres which arise in the meso- thoracic ganglion and pass into the median nerve and others are innervated from the metathoracic ganglion; the rest of the mesothoracic musculature is innervated through the lateral nerve. A similar pattern is repeated in the metathorax. It is interesting to note that two of the middle bands of fibres in the dorsal longitudinal muscle receive a double innervation; apart

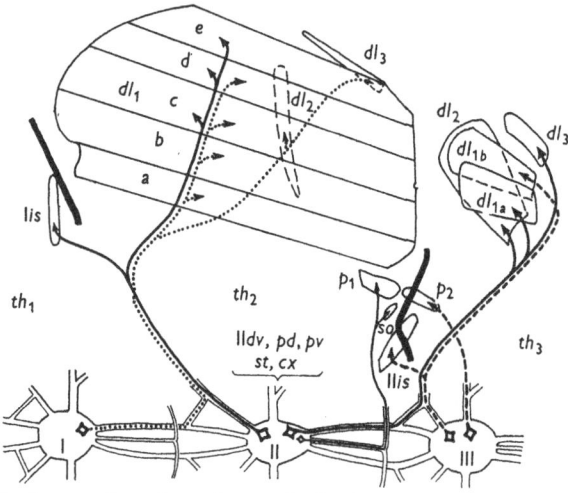

Fig. 40. The innervation of certain muscles in the meso- and metathorax of *Telea polyphemus; cx,* coxal; *dl,* dorsal longitudinal; *dv,* dorsoventral; *is,* intersegmental; *p,* pleural; *pd,* tergopleural; *pv,* pleurosternal; *st,* sternal; *th,* thoracic segments. (From Nuesch, 1954, corrected after Nuesch (unpublished)).

from a brief mention by Mangold (1905) of *Dytiscus* flight muscle, the only other report of this feature, which is normal in leg muscles, is a statement by Tiegs (1955) that the dorsal longi- tudinal muscles of *Cyclochila australasiae* (Homoptera, Cicadidae) and *Erythroneura ix* (Homoptera, Jassidae) have two separate nerve fibres in the final branches that go to form the motor end-organ; in the same muscle in muscid Diptera there is only a single motor fibre (Pringle, 1949).

Otherwise the evidence about motor innervation is indirect. In insects with a neurogenic rhythm (indirect muscles of the 1:1 type of Roeder, 1951), separate motor fibres to the various flight muscles would seem to be required for control of wing movements (Weis-Fogh, 1956*b*). The pattern of electrical spikes recorded from the thorax during flight gives some information. The record of fig. 27 c suggests that there are impulse volleys in different muscles at four different instants in the wing-beat cycle of a sphingid moth; Pringle (1949) showed that, during the steady 'anaesthetic flight', there is a regular discharge at about 3 impulses per sec. in at least five motor fibres supplying different parts of the (myogenic) indirect muscle complex in *Calliphora*. The fact that flight is possible in Diptera after decapitation shows that these fibres must originate in the composite thoracic ganglion. In *Periplaneta americana* Chadwick (1953) reports that flight movements never occur if both connectives are cut between the prothoracic and mesothoracic ganglia, and that the hind wings are not moved if both connectives are cut between the meso- and metathorax. This suggests that in this insect, as in *Telea polyphemus*, some of the motor fibres to the flight muscles arise in the ganglion of the segment in front.

SENSE ORGANS IN THE WINGS. The wings of all insects which have been examined receive a sensory nerve trunk, fibres of which run in the main veins to some distance out along the wing. The nature and distribution of the sense organs from which these fibres originate have been studied in detail by (besides earlier workers) Vogel (1911, 1912), Lehr (1914), Erhardt (1916), Eggers (1928), Hertweck (1931) and Zaćwili-chowski (1933 a–c, 1934 a–e, 1936 a, b). The organs are of three main types: bristles and hairs of various lengths, campaniform sensilla and chordotonal sensilla. The innervated bristles and hairs are widely distributed in certain orders, with small sensory cell bodies and small-diameter sensory fibres. Since insects respond to a light touch on the folded wings, they are probably mainly tactile in function, but it cannot be excluded that they play a role in flight by reacting to the air flow over the moving wings; the small diameter of the nerve fibres supplying them (and correspondingly slow impulse conduction) suggest, however, that they do not mediate the rapid reflex actions necessary for the control of flight. The sensilla of the other two types have a precise form and location in any given species and there are indications of a common plan of arrangement throughout the class; since their mode of action is known from neurophysiological studies of similar organs elsewhere in the body, it is possible to make a tentative appraisal of their role in flight.

Campaniform sensilla consist of a single large sensory cell

whose distal process is inserted in the centre of a domed region of thin modified cuticle; the sensory axons are of large diameter. In external view the sensillum is circular or oval in outline and the dome may be level with or sunk below the general surface of the cuticle. Campaniform sensilla are mechanoceptors, sensitive to strains in the exoskeletal sheet (Pringle, 1938a). They often occur in groups, and the specialized, oval type shows a constant orientation within the group of the long axis of the oval; such groups form a unitary organ selectively sensitive to strains whose compression component is orientated parallel to the direction of the long axis. The leg campaniform sensilla of *Periplaneta* are known to function as proprioceptors, reflexly controlling muscle tension in relation to the force exerted by the leg on the ground (Pringle, 1940).

Chordotonal sensilla have large, elongated sense cells which are attached internally between two points on the cuticle. They commonly occur in compact bundles of cells numbering from three or four to several hundred. The largest and best-known groups occur in the tympanal and sub-genual organs (Eggers, 1928), where they are sensitive to vibration (Pumphrey and Rawdon-Smith, 1936; Autrum and Schneider, 1948), but sensilla have also been described (Hughes, 1952) with slow adaptation to a constant stimulus. In all cases they appear to register the changes in length between their two points of attachment, and it may be inferred from their structure that extension rather than compression is the adequate stimulus.

Table 3 summarizes the results of the workers listed above on the distribution of campaniform and chordotonal sensilla on the wings. The classification is based as far as possible on the conclusions of Zaćwilichowski, who has tried to establish a common pattern of grouping, but, owing to variation in the venation of different orders and the absence from the literature of good discriminatory criteria when one group merges into another, detailed homologies are uncertain in some cases. In addition to the groups of campaniform sensilla shown in the table, there are usually scattered sensilla elsewhere on the wing veins; the numerous large sensilla on the lower surface at the wing margin in Lepidoptera are particularly noteworthy (Vogel, 1911). It is most unfortunate that none of the histologists studying campaniform sensilla has paid attention to the orientation of the long axis in the oval type, since this information is essential to an

TABLE 3. *Distribution of campaniform and chordotonal sensilla on the wings of insects. The more distal, scattered campaniform sensilla are not listed. The figures show the numbers of sensilla in each group, or the total number when the groups have not been distinguished. An asterisk signifies that the group is present and a dash that it is absent.*

Order or family	Species	Wing	Campaniform groups							Chordotonal organs				Author
			Lower surface			Upper surface				Ante-alar	Radial	Medial	Cubital	
			Sc1	Sc2	Sc3	R1	R2	R3	A					
Blattidae	*Blattella germanica*	Fore	5–7		6–9	—	—	—	—	6–8	6	—	—}	Zaćwilichowski (1934a)
		Hind	5–7		—	—	—	—	—		6	—	—}	
Acrididae	*Chorthippus biguttulus*	Fore	5–10		19–23	—	—	—	—	—	—	—	—}	Zaćwilichowski (1934b)
		Hind	5–		17–20	—	—	—	—	—	—	—	—}	
	Tettigonia (Locusta) cantans	Hind	10–12		—	—	10–12	—	—	—	—	—	—	Erhardt (1916)
Plecoptera	*Chloroperla (Isopteryx) tripunctata*	Fore	8		3–4	—	—	—	—	4	—	—	—}	Zaćwilichowski (1936b)
		Hind	6–7		5–6	—	—	—	—	4	—	—	—}	
Odonata	*Coenagrion puella*	Fore		25		60 radial, 40 costal			*	—	—	—	*	Erhardt (1916)
Homoptera	*Aphrophora alni*	Fore	8–9		5–7	18–19		7–11	2–3	10–12	14–16	—	—	Zaćwilichowski (1936a)
		Hind	___		7–10		40				16	—	—	
Neuroptera	*Chrysopa carnea (vulgaris)*	Fore		*			*			*	*	*	*	Erhardt (1916)
		Hind		*			*			*	*	*	*}	
Mecoptera	*Panorpa communis*	Fore	10	7–12	7–12	19–23	46–53	5–7	—	6–12	20–24	8–12	6–10}	Zaćwilichowski (1933a)
		Hind	17–25		11–12				—	8–12	*	*	*	
Trichoptera	*Anabolia laevis*	Fore		6–7			120		*		8–12	*	4–6}	Zaćwilichowski (1933c)
		Hind		8–12		75–90		9–12	*		8–12	*	*	
Lepidoptera	*Cerura vinula*	Fore	80	18		80	9	82	9	?	?	?	?}	Vogel (1911)
		Hind		18		160	50	22		?	?	?	?	
Coleoptera	*Dytiscus marginalis*	Elytron	—		—	130–150	130–150			—	30–40	—	—}	Lehr (1914)
		Hind		30		300–400	300–400			—		—	—	
Hymenoptera	*Apis mellifera*	Fore	350–400	30–50	50	250	5–7	5–9	30	16–20	24–30	10–12	*}	Zaćwilichowski (1933b)
		Hind	325–350	30			70–80		6–7					
	Vespa rufa	Fore	300	70	9	170		8	?	—	*	—	—	Erhardt (1916)
Diptera	*Tipula paludosa*	Fore	21–22	14–18	9–12	50–60	30–36	6–10		4–6 (4–6)²	16–20 (3–4)³	—	4–6}	Zaćwilichowski (1934d,e)
		Haltere	8–9	40–50	4–5	5–10	56–78	60				—	6	
Pupipara	*Crataerina (Oxypterum)* sp.	Fore	7	3–6	7	10–13	5–6	2		6–7	6–7	—	—	Zaćwilichowski (1934c)

understanding of their role in flight. A few wings have been re-examined with this in mind.

Generally, and perhaps universally, the main groups of campaniform sensilla are found on the lower surface of the wing on the sub-costa and on the upper surface on the radius, and for convenience of presentation all the main upper-surface groups are shown in table 3 as radial and all the main lower-surface groups as sub-costal, on the assumption that workers such as Lehr (1914), who located them differently, were in error in the

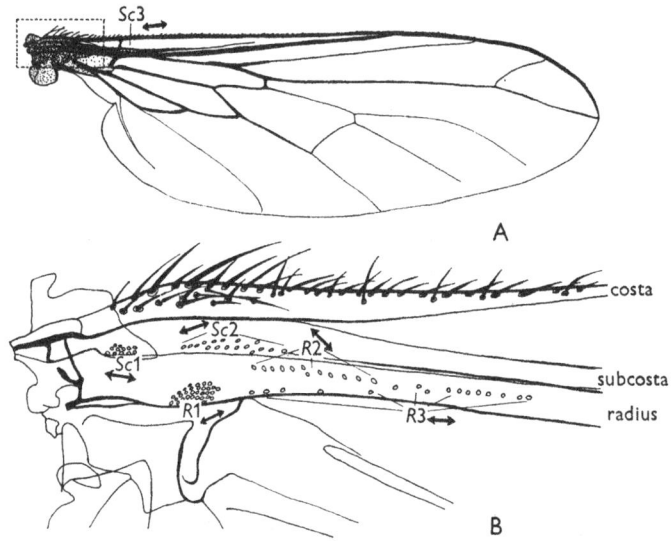

Fig. 41. A, right wing of *Empis tessellata*, B, enlarged view of wing-base (inset in A), to show the location and orientation of the groups of campaniform sensilla. R_1, R_2, R_3, radial groups; Sc_1, Sc_2, Sc_3, subcostal groups, with arrows indicating the orientation of the long axis of the sensilla. (Original; *ex. coll.* J. Smart.)

naming of veins. The following original observations have been made on the structure and orientation of these sensilla. The scattered sensilla are always large and circular in outline (also Vogel, 1911, who gave their internal diameter as 13μ in *Erannis* (*Hibernia*) *defoliaria*). In powerful, efficient fliers such as Diptera, Hymenoptera and Coleoptera, the grouped sensilla are always oval, and the size of the grouped sensilla is always less than that of the scattered sensilla ($4\cdot5–9\mu$ in *Erannis*; Vogel, 1911). The orientation of the long axis differs in the different sensilla groups, but is constant for the homologous group in all species in which it could be determined; it is shown in fig. 41 for *Empis tessellata* (Diptera), where the orientation is readily observed and whose sensilla groups can be precisely homologized in the classification of Zaćwilichowski.

The location of the chordotonal organs is shown in fig. 42 for *Panorpa communis*, in which all four are present. The ante-alar organ is inserted distally on the lower membrane of the base of the wing at the costal margin; the radial organ (and the medial when present) stretches from the proximal, upper posterior surface to the distal, lower anterior edge of the vein; the cubital

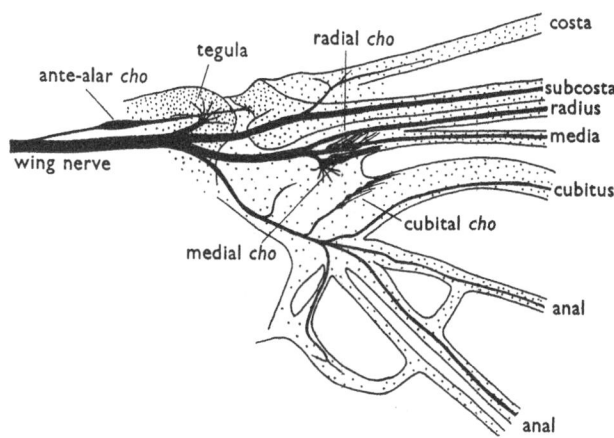

Fig. 42. Right wing-base of *Panorpa communis*, to show the location of chordotonal organs (*cho*). (Redrawn from Zaćwilichowski, 1933a.)

organ also has its distal terminal insertions on the ventral surface. These orientations are remarkably constant in different orders and suggest a true homology.

The absence of chordotonal organs from the wings of Orthoptera has been noted both by Erhardt (1916) and Zaćwilichowski (1934b). The basal groups of campaniform sensilla on the upper surface of the radial vein are absent in Blattidae, Acrididae and Plecoptera and appear to be replaced by thick-rimmed hair sensilla which may have a similar function (Zaćwilichowski, 1936b).

An elaborate development of both campaniform and chordotonal sensilla occurs in the halteres of Diptera (Pflugstaedt, 1912; Zaćwilichowski, 1934e) (fig. 43); Zaćwilichowski establishes homologies with the wing groups as shown in table 3. The ante-alar chordotonal organ is double and the radial organ triple; the terminal processes of the chordotonal sense cells also tend to spread out over a wide area of cuticle in certain wings.

Before attempting to assign particular roles to the components of this elaborate array of sensory structures, it is useful to consider the types of mechanical stimulus which are available for monitoring the movements of flight. We saw in Chapter 4 that in the dynamics of wing motion aerodynamic, inertial and elastic torques all play significant parts. It is noteworthy that all

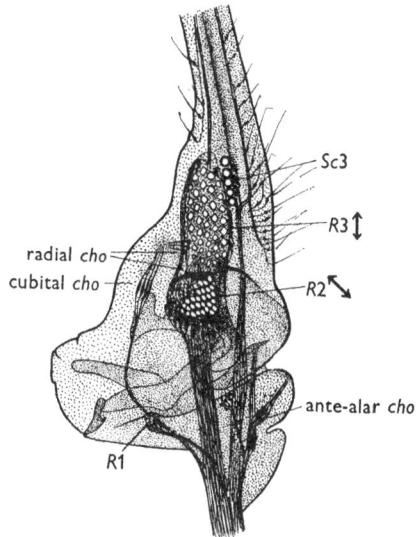

Fig. 43. Dorsal view of the base of the left haltere of *Tipula paludosa*, to show the location and orientation of the groups of campaniform and chordotonal sensilla: *cho*, chordotonal organs; $R1$, $R2$, $R3$, radial groups, $Sc3$, subcostal group of campaniform sensilla, with arrows indicating the orientation of the long axis of the sensilla. (Redrawn from Zaćwilichowski, 1934e, with orientation added from direct observation.)

the wing sense organs with the possible exception of the ante-alar chordotonal organ are situated distal to the hinge which contains the elastic elements of the flight system. Only aerodynamic and inertial torques need therefore be taken into account in a discussion of the strains and distortions in the veins bearing the sense organs. We have knowledge of the magnitude and time course of these two torques only for the flight of *Schistocerca* (Jensen, 1956), but it is instructive to use these results to plot curves showing the magnitude and direction of action of the combined aerodynamic and inertial torques at the base of the fore and hind wings of this insect (fig. 44). These curves show the bending moment which would be present through the wing-beat cycle if the entire load of the wings was carried on a single cylindrical rod pivoted at the fulcrum; such a simple articulation is not, of course, found, and the torque is in fact transmitted through the whole cuticle, but by far the greater part of the load must be carried by the main basal veins, on or in which lie the chief groups of campaniform sensilla and the radial chordo-

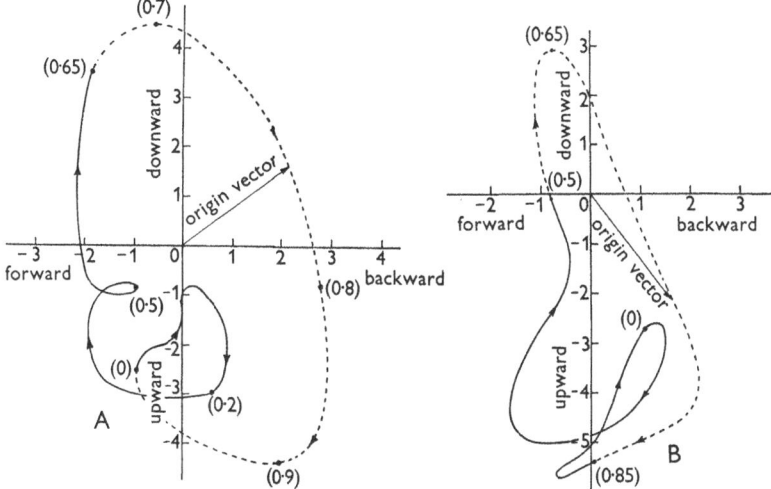

Fig. 44. Graphs showing by the origin vector the magnitude in g. cm. and the direction of action of the torque in the basal veins of the fore wing (A) and hind wing (B) of *Schistocerca* during one stroke; solid sections indicate downstroke, broken sections indicate upstroke; the figures in brackets show fractional instants during the cycle, starting from the beginning of the downstroke of the fore wing. (From data in figs. III, 22 and III, 23 of Jensen, 1956.)

tonal organ. It should also be remembered that in Orthoptera, as in Odonata, Coleoptera and probably Lepidoptera and other orders with large direct muscles, part of the motive power for wing movements is transmitted not through the basal articulation but by apodemes inserted distal to some of the groups of sensilla.

The most noteworthy features of the curves of fig. 44 are as follows:

(i) In both wings the maximum torques act downward and slightly forward, upward and slightly backward. This achieves the motion in the stroke plane.

(ii) At different instants in the wing-beat cycle the torque at the wing base acts in different directions; the most rapid change of the torque vector occurs during the upstroke ($0·7$–$0·9$ in the fore wings; $0·65$–$0·85$ in the hind wings).

(iii) The double-loop appearance of both diagrams is due to a torque component oscillating at twice the wing-beat frequency. This $2n$ component comprises a larger fraction of the total in the direction at right angles to the stroke plane than in the stroke plane, particularly in the hind wings.

The wing motion of all insects is sufficiently similar to that of *Schistocerca* for it to be reasonable to assume that diagrams of generally similar shape to those of fig. 44 would be found for other species, though in detail the curves will depend on the form of the stroke. Pringle (1948) showed from a consideration of the dynamics of haltere motion that a complicated cycle of torque changes at the base is also generated in this organ due to gyroscopic forces arising during rotation of the whole insect.

We may now consider the probable mode of action of the sense organs. Each of the groups of campaniform sensilla should be maximally sensitive to a particular direction of torque in the vein in which it lies. The groups on the lower surface of the sub-costa and on the upper surface of the radius are located in the positions in which the maximum strain occurs. Those with orientation longitudinal to the wing or haltere axis should indicate primarily torques in the stroke plane; those with diagonal orientation may be expected to react to other directions of torque including the component at right angles to the stroke plane. Well-differentiated elliptical sensilla should be more selective in their response than circular sensilla and, if their sensitivity is correctly adjusted, might respond only once per wing stroke; the groups would therefore deliver volleys of impulses at different instants in the cycle. In the case of the haltere where the orientation of the sensilla is extremely clearly defined, Pringle (1948) made the specific suggestion that those of the basal plate (R_2, fig. 43) were responsible for the selective indication of gyroscopic torques developed during rotations of the fly; on the wings, where there is a greater variety of orientations in the different groups (fig. 41) the potential exists for the registering of a large amount of proprioceptive information about the mechanical result of the wing beats.

The figures for sensilla distribution in table 3 contain two pieces of evidence consistent with this view. There is, first, a broad correlation between the number of sensilla in the groups and the manoeuvrability of the species in flight. It is generally true that the possession of a large number of endings of similar qualitative sensitivity but varying threshold confers a greater power of discrimination of the stimulus intensity, and in the particular case of campaniform sensilla Pringle (1938) showed that a numerically large group with regular arrangement should also have better qualitative discrimination. Secondly, it becomes significant that the elytra of beetles have no sensilla on the lower surface, although the single upper-surface group is numerous. The torque during flight at the base of the non-oscillating elytron must be derived from the relatively constant aerodynamic forces, and a qualitatively single sensory indication with good quantitative discrimination is all that is required to account for observations such as that of Demoll (1918) that the speed of flight is reflexly doubled in *Melolontha* by removal of 45 % of the elytra. The constant orientation of the sensilla in the elytral group has been confirmed in *Cantharis rusticus*, which is as good a flier as any of our English beetles.

The mode of action of the chordotonal organs is less easy to deduce from their anatomical arrangement. It is to be expected that the torques at the wing base will produce some distortion of the cuticle which might be indicated by suspended sensilla reacting to an increase in length; Zaćwilichowski (1934a) remarks that, in close proximity to all insect chordotonal organs, there is always found a region of very thin, wavy cuticular membrane, which must tend to concentrate displacements in this region in comparison with the more heavily sclerotized regions bearing the campaniform sensilla. It cannot be excluded that some of the wing chordotonal groups in certain orders may have an auditory function; Vogel's (1912) conclusion to this effect is based not only on the analogy with known tympanal organs but also on the presence of tympanum-like membranes at the point of attachment of the sensilla in some Lepidoptera. Other possibilities are that the changes of air pressure during the wing stroke may be directly monitored by tympanum-like membranes if the rate of adaptation of the sensilla is slow. In many orders, however, none of these modes of operation is indicated by the anatomy, and a straightforward registration of the distortion produced by the wing motion seems probable.

On this basis the typical arrangement such as is found in *Panorpa* (fig. 42) suggests that the radial and medial organs should signal motion in the stroke plane and the ante-alar and cubital organs movements in the plane at right angles, including perhaps the movements of folding and unfolding of the wings. Such a conclusion is consistent with the suggestion of Pringle (1948) that the large chordotonal organ of the muscid haltere (the cubital organ in the scheme of Zaćwilichowski; fig. 43) contributes to the perception of the gyroscopic torques; the haltere ante-alar organ, which was not mentioned by Pflugstaedt (1912), is suitably situated to play a similar role.

In conclusion to this section it should be noted that no muscle receptor sense organs have ever been described from any of the flight muscles of insects. Internal receptors similar to those in the muscles of Crustacea (Alexandrowicz, 1951; Wiersma, Furshpan and Florey, 1953) have been found in the dorsal longitudinal muscle of meso- and metathorax and abdominal segments of a lepidopteran larva, and in the abdomen of pupa and adult, but not in the adult thorax (Finlayson and Lowenstein, 1955); also in the abdomen of various Acrididae and in the abdomen and thorax of the flightless stick-insect, *Carausius morosus* (Slifer and Finlayson, 1956). Until thoracic muscle receptor organs have been shown to exist in a winged insect it must be assumed that direct proprioceptive indication of flight movements is provided entirely by sense organs on the wings themselves.

OTHER SENSE ORGANS. Flight may be initiated and to some extent affected in detail by stimuli from many parts of the body, but two sense organs in particular have been shown to be specifically concerned with flight, though located far from the wings. These are the antennal mechanoceptors, and the frontal hair plates.

The importance of the antenna in flight regulation was first shown by Hollick (1940). A jet of air impinging on the antennae of *Muscina stabulans* (Diptera) so as to produce rotation of the third joint and its arista relative to the second joint caused the legs to be drawn up into the flight attitude. Covering the antenna with a small cap produced a condition in which flight could be initiated in a mounted fly by removal of contact with the legs, but not maintained in an air stream. In some insects this stimulatory effect of the air flow on the antennae became unnecessary for maintained flight after 24 hr., but there remained a marked difference in the wing-tip path in an air stream between intact and antenna-less flies. Hollick suggested that the sensory element in all these reactions was Johnston's organ, a complex group of chordotonal sensilla which occurs in the correct location in most insect orders and with an arrangement well suited to register the relative movement of the antennal joints (Eggers, 1928). Electrophysiological studies are required to determine whether the adequate stimulus to this

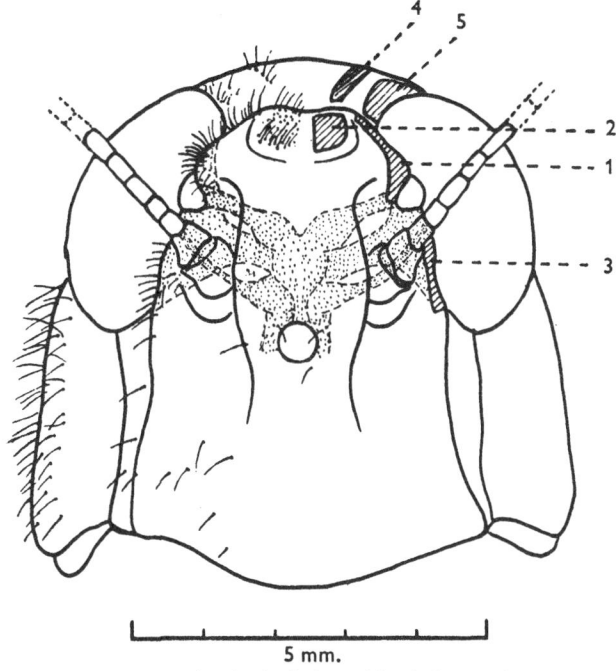

Fig. 45. The five pairs (1–5) of wind-sensitive hairs on the upper part of the head of *Schistocerca gregaria*. (From Weis-Fogh, 1956b.)

sense organ is the steady deflexion produced by air flow or the vibration which will probably also be present.

The frontal hair plates were first described by Weis-Fogh (1949) from the locusts, *Schistocerca gregaria* and *Locusta migratoria*. In *Schistocerca* they comprise five patches of short, dense hair sensilla on each side of the frons and vertex of the head (fig. 45), innervated by a special dorso-tegumentary branch of the paired cerebral nerves (Albrecht, 1953) and homologous with

Eltringham's organ of Lepidoptera (Eltringham 1925; Ehnbom, 1948). Stimulation of these areas with a fine air jet sufficed to bring the fore legs (but not the hind legs) into the normal flight attitude and to induce and maintain wing movements; stimulation of the antennae or of other parts of these insects was ineffective. Weis-Fogh (1949) suggested that a steady deflexion of the hairs in the air flow was the adequate stimulus, but he later (1956 b) inclined to the view that vibrations were more important since static deflexion by a thin needle produced no response; electrophysiological studies on similar hair plates on the legs of *Periplaneta* (Pringle, 1938 c) demonstrated that steady deflexion can be an adequate stimulus, but that the sensitivity to vibration is extremely high. Weis-Fogh (1949) also showed that a jet of air impinging on the lateral frontal hair plate (3 in fig. 45) from one side produced in a suspended flying locust a turning movement by asymmetrical wing action; but he has not established by wind-tunnel experiments that this sense organ is involved in a yaw-stabilizing reflex in normal flight (Weis-Fogh, 1956 b).

It has been claimed (Jawlowsky, 1936) that the fibres of the dorso-tegumentary nerve pass through the tritocerebrum in Coleoptera without synapsing, and enter the oesophageal connectives. Burtt and Catton (unpublished) have found in *Schistocerca* that impulses apparently arising from the frontal hair plates can be picked up electrically in the connectives. If these observations are confirmed they provide strong evidence for a special function for these groups of sensilla.

REFLEXES AND CENTRAL NERVOUS MECHANISMS. Flight is one of the most complicated forms of locomotion found in the animal kingdom and its regulation demands great precision of nervous control. Perhaps because of the close integration of the many factors involved in this process, our knowledge of it is still rudimentary. As always in physiology, central nervous phenomena cannot be analysed without an understanding of the sensory input and motor output channels; only recently has sufficient progress been made with such studies in insects to make possible a critical study of flight reflexes.

For instance, there are many examples in the literature (see Chadwick, 1953) of investigations of the influence of various factors on the wing-beat frequency of Diptera and on other parameters of the flight system. It is now apparent that some of these factors (e.g. inertia, wing-loading, temperature) may exert their influence directly on the myogenic muscular machine, either without the involvement of reflexes or with reflex effects superimposed. Since not all insects have a similar motor mechanism, the same over-all effect in different orders may be evidence of different physiological components. Unknown pitfalls may lie hidden on the sensory side.

It is clear, however, that in the flight system we have to deal with a well-circumscribed problem. To be effective, flight demands a certain minimum of activity and there is rarely any doubt about whether the system is in action or not. Because of this feature, a distinction can be drawn between the reactions of flight initiation and those of flight maintenance; it is probably also valid to treat separately the reactions to mechanical and other influences which comprise the mechanism of co-ordination, and the more general reactions to environmental stimuli by which flight is made to play an effective role in the life of the insect. These last form part of the study of insect behaviour and will not be discussed here in any detail.

INITIATION AND MAINTENANCE. Many forms of noxious stimulation will initiate flight in insects which use this means of escaping from danger, but the most important and widespread specific response is the so-called 'tarsal reflex' of Fraenkel (1932). In nearly all insects tested, including *Periplaneta*, *Schistocerca*, various Odonata, *Macroglossum*, a cetoniid beetle, *Vespa*, *Apis* and *Calliphora*, removal of a contact stimulus to the legs at once initiated wing movements, and establishment of leg contact at once inhibited them. The only exception found was *Melolontha* which, if its wings were folded, either showed no flight reaction or else flew only after going through the elaborate preparatory operations characteristic of beetles; wing movements were, however, at once inhibited by leg contact. Fraenkel found that this flight-inhibiting reflex from leg receptors was so important in *Macroglossum* and *Calliphora* that a legless insect might be incapable of stopping its wing movements; he states specifically for *Vespa* that removal of the tarsi had the same effect as removing the whole of the legs, and concludes that the receptors must lie on this joint, but in view of the finding of Pringle (1938a, 1940) that it is the leg campaniform sensilla on the trochanter and femur which are concerned in the inhibition of the 'Suchenreflex' in *Periplaneta*, the point needs reinvestigation.

The duration of the period of wing movement which may be

initiated through this reflex in a suspended insect varies in different species. *Drosophila* (Chadwick, 1939; Wigglesworth, 1949) may continue flying until exhausted without further exteroceptive stimulation; *Macroglossum* (Fraenkel, 1932) may continue for half an hour. In most insects wing movements initiated in this way usually cease after a period of seconds or minutes unless reinforced by stimuli for flight maintenance. These seem to be mainly of two types. Air flow on the head is influential in *Muscina* (Hollick, 1940) and *Schistocerca* (Weis-Fogh, 1949); the nature of the sense organs concerned has already been discussed. In addition, Weis-Fogh (1956*b*) finds that air flow (from a wind tunnel) is an adequate flight-maintenance stimulus in *Schistocerca* even when the frontal hair plates have been immobilized or the brain removed. He calls this stimulus 'wind on the moving wings' and attributes the response to sense organs monitoring the aerodynamic forces, possibly specifically the lift on the upstroke; only this stimulus suffices to maintain the hind legs in normal flight posture. In this respect *Schistocerca* appears to differ from *Muscina*, in which Hollick (1940) found wing movements to be of short duration even in an air stream if the antennae were removed; also, air flow on the antenna brought all the legs of *Muscina* into complete flight posture. Weis-Fogh argues that this purely thoracic flight-maintenance reflex is a primitive one because of its operation in decerebrate locusts which have very regular wing motion; its existence has several times been suggested on general grounds (von Buddenbrock, 1919; Fraenkel, 1932), but has not before been conclusively demonstrated.

Other types of flight-initiating reflex have been described, notably a response in *Periplaneta* to change in the relative position of pro- and mesothorax (Diakonoff, 1936). There is also clearly a close connexion in many insects between jumping and the initiation of flight (Fraenkel, 1932), but it is difficult to separate an inherent connexion between these two responses from the loss-of-contact stimulus to the legs which inevitably results from the jump. A possible anatomical link between jumping and flight initiation was mentioned on p. 38; in a wide range of insect families the same tergo-trochanteral muscle could be used for the two purposes (Smart, 1957).

CO-ORDINATION. The integration of activity in many muscles which is necessary for effective flight is probably brought about partly by the innate organization of the nervous centres in the thoracic ganglia. A movement of this sort, however, requires to be so closely regulated and adjusted that reflex effects from sense organs reacting to the results of the movements can be expected to be present in all insects. Since the plan of the arthropod nervous system does not permit de-afferentiation to be carried out by nerve section as in the vertebrates (Lissmann, 1946), the existence of innate patterns of central nervous organization can only be inferred after exclusion of reflex explanations of any given feature, and it is convenient first to discuss regulatory reflexes. These are arranged as far as possible in terms of the flight parameter affected, although the nature of the experimental evidence is rarely such as to facilitate such a rigorous classification. An attempt to display a total picture of flight regulation is made in fig. 52.

TABLE 4. *The effect of altering the body angle on various flight parameters in steadily flying locusts* (from Weis-Fogh, 1956 b)

Body angle (degrees)	0	5	10	15
Mean relative lift (%)	102	105	102	94
Mean wing-beat frequency (min.$^{-1}$)	1080	1080	1080	1090
Mean flying speed (m./sec.)	3·8	3·4	3·0	2·6
Mean stroke-plane angles (degrees):				
Fore wing	31	34	36	42
Hind wing	29	31	33	36
Mean stroke angles (degrees):				
Fore wing	65	69	71	80
Hind wing	110	110	111	113

LIFT. Weis-Fogh (1956 a) has made a statistical survey of the variations in the flight system of tethered locusts flying in a wind tunnel, and (1956 b) has studied in particular the changes which occur when the angle of the body to the air flow is altered. In his experiments the air velocity from the wind tunnel was always adjusted so that there was no horizontal force acting on the suspended flying insect (aerodynamic thrust = body drag). He found that, over a range of body angles from 0° to +15° (head up) the lift was remarkably constant (table 4) at a mean value close to the basic weight (the weight of a typical locust after 1 hr. of flight). In a given individual the wing-beat frequency changed hardly at all, and the change in other measurable parameters such as mean flying speed, stroke plane and stroke amplitude could not account for this regulation of the lift. Weis-Fogh concluded that lift is regulated mainly by a reflexly controlled pronation twist of approximately the same amount as the angular increase in body incidence. The twist is of importance chiefly during the downstroke and involves a change in the relative excitation or phasing of excitation to the basalar and subalar direct muscles; passive or purely muscular regulation is not adequate to account for the effect. There is as yet no evidence on the nature of the sensory component of this lift-regulating reflex, but from its characteristics it seems probable that it is distinct from the 'wind-on-the-moving-wings' flight-maintenance response discussed on p. 102.

A similar conclusion, that alteration of the angle of attack of the wings is the main factor concerned in the regulation of lift, was reached by Chadwick (1953) through an elaborate argument based on experiments with *Drosophila* spp. The significant observations (Chadwick and Williams, 1949; Chadwick, 1951) are (i) that wing-beat frequency and stroke amplitude both vary in an inverse exponential relationship with air density to which aerodynamic forces are directly proportional; (ii) that the rate of oxygen consumption is unaffected by a change in pressure from 760 to 200 mm. Hg in pure oxygen. Both sets of experiments were done with suspended flying insects. Assuming that the 'efficiency' of flight is unaffected by a change in pressure, Chadwick concludes from (ii) that the total power output is constant, and that since power output depends only on air density, stroke amplitude, wing-beat frequency and some function of the angle of attack, and the observed relationship between the first three parameters does not give a constant power output, there is, by exclusion, only the angle of attack left to be adjusted in order to account for the regulation. The argument does not depend on the misleading function (sin α) chosen for relating power output to the angle of attack (see p. 69), but is nevertheless dubious because of the big assumption of constant 'efficiency'. The conclusion, however, is the same as that reached by Weis-Fogh (1956*b*) and seems likely to be correct. The mechanism of control of wing-twisting in a fly is, of course, different from that in the locust, but again involves some of the direct muscles. There is no reason why the same type of lift-regulating reflex should not have been preserved through the whole evolution from a neurogenic to a myogenic system of power generation.

FREQUENCY. As has already been pointed out, many of the factors known to affect wing-beat frequency in Diptera and other orders with fibrillar muscle are now seen to operate by a direct action on the muscles rather than through a reflex; of these, wing inertia is the most significant (Pringle 1949; Roeder, 1951; Sotavalta, 1952). The true measure of the reflex effect of changes in wing inertia is the frequency of motor-nerve impulses to the flight muscles, which is accompanied by a change in wing-beat frequency only in species with the synchronous (1:1) mechanism. On the importance of such reflexes there is disagreement. Roeder (1951) showed that in *Periplaneta* wing-cutting produces little change in wing-beat frequency and in *Agrotis* (Lepidoptera) a slight decrease. On the other hand, Sotavalta (1954b) and Tiegs (1955) found that in many Lepidoptera and in some Neuroptera and Isoptera (all with non-fibrillar muscle) a reduction in inertia produces a slight increase in frequency, though of a smaller order of magnitude than in the orders possessing fibrillar muscle; Sotavalta confirmed the absence of effect in *Periplaneta*. In *Vespa* and *Calliphora* (with fibrillar muscles) Roeder (1951) found that amputation produced a decrease in the frequency of motor-nerve impulses. Taken over-all, the results indicate that there is present in most insects a reflex relating wing inertia to impulse frequency in the flight motor nerves, but that its operation is variable and its influence insufficient to produce any effective regulation of power output. Weis-Fogh's (1956a, b) demonstration that frequency in *Schistocerca* depends more on the size of the individual insect than on any variable which can change in a particular individual lends support to the view that frequency is not an important component of the reflex mechanism of flight regulation; its appearance as a dominating factor in the myogenic mechanisms is probably more an inevitable result of the physiology of the muscles than a feature of functional importance.

The sense organs responsible for the reflex have not been determined; they might be the same as those concerned in flight maintenance, and the whole response merely a corollary of this maintenance system.

STROKE AMPLITUDE AND POSITION. A different type of regulation was studied by Hollick (1940) in the flight of *Muscina*, involving control, not of the magnitude of the total lift force, but of the line of action of the aerodynamic resultant in relation to the centre of gravity. Such studies are relevant to the problem of stability in the free-flying insect, since an average turning couple will be present if the average aerodynamic force does not act through the centre of gravity. It is important here to appreciate a point of general principle (Pringle, 1950). Any flying machine, to be controllable in the air, must possess stability. By this is meant a tendency to return to a certain attitude when slightly displaced. It is possible to conceive theoretically of a flying machine with active (reflex) stabilization in all the three planes of rotation, but owing to the interaction of the different movements in a heavier-than-air machine the integration of the reflexes would have to be very elaborate. In all man-made flying machines the aerodynamic design ensures that stability in at least two of the three planes is achieved by a stable shift of the centre of pressure, so that a turning couple is automatically produced tending to restore the original attitude after angular displacement. When such inherent stability is present, it is still necessary to have *control*, in the sense of an adjustment of the equilibrium point of the system, but this is a different thing from active *stabilization* and requires less complicated feed-back pathways.

Hollick's (1940) experiments were concerned with stability and control in pitch (rotations in the fore-and-aft plane about a transverse axis) and with the regulation of air speed. He found that in thirty-two insects performing regular wing movements when fixed from the dorsum to a flight balance the steady aerodynamic force was 30–35 mg. acting, on an average, along a line inclined forwards and upwards at an angle of 48° to the body axis of the fly (average weight of flies, 31 mg.) (fig. 46B). This force line, however, did not usually pass through the centre of gravity of the insect, being well behind it in the majority of cases; there was, therefore, a forward pitching moment. Simultaneous observation of the stroke amplitude revealed a correlation in the population between this flight parameter and the line of action of the aerodynamic resultant, the smaller stroke amplitudes producing a larger forward pitching moment (fig. 46A). The reality of the pitching moment was confirmed by the rapid change of body axis which occurred when flies with a stroke amplitude of 140–150° were suddenly released. He next determined the effect of air flow from a wind tunnel on the line of action of the resultant in flies held with their body axis at various inclinations to the horizontal. Increase in air flow produced a forward movement of the line of action, the magnitude of the change being greater with the body axis horizontal than when inclined upwards; the combined effect of the three variables was such as to produce a flight system dynamically stable in pitch, the equilibrium point being with the body axis inclined upwards at 15–29° and with a forward motion at 160–230 cm. per sec.

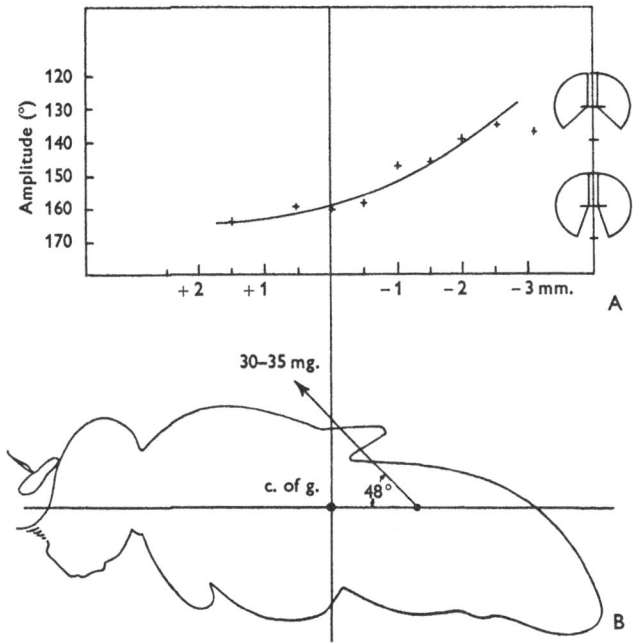

Fig. 46. Experiments with *Muscina stabulans* (Diptera). A, the relationship between stroke amplitude and the horizontal distance from the centre of gravity to the line of action of the aerodynamic resultant. B, diagram showing the usual magnitude and line of action of the resultant in a suspended insect flying in still air. (Redrawn with additions from Hollick, 1940.)

It is noteworthy that Hollick found little or no change in the stroke amplitude or the line of action of the aerodynamic resultant when the body angle of his flies was altered in still air; he thus had no good evidence for a static gravity sense. A change in the pattern of wing movements during static tilts of ±45° in the pitching and rolling planes has been claimed by Schneider (1953) from experiments in which the downflow of air from a tethered flying *Calliphora* was indicated by the pattern of displacement of lycopodium powder on a card. The effects were, however, also shown to a lesser extent in anaesthetic flight, and unfortunately figures are not given of the magnitude of the change in the normal insect and the control. In view of the negative evidence of Faust (1952) further observations are needed before it can be regarded as established that there is a static sense in muscid Diptera.

THE HALTERES OF DIPTERA. That the halteres are of great importance in the regulation of flight in many Diptera has been clear since Derham (1713) showed that flies deprived of these organs are unable to maintain balance. Experimental proof of this function for the halteres was given by Fraenkel (1939), and the mode of action of the muscid haltere as an organ of special sense was worked out by Pringle (1948), who established that the gyroscopic properties of the oscillating mass generate forces at the base of the stalk during rotations of the whole fly which can be perceived by the many groups of campaniform and chordotonal sensilla located in this region. On theoretical grounds and on the basis of some preliminary flash-photographs of a free-flying, haltere-less fly (*Eristalis tenax*), Pringle suggested that the halteres are responsible particularly for reflex stabilization in the yawing plane.

Since that date two important papers have appeared dealing with the actual regulatory mechanisms, and it is now clear that the role of the halteres is wider than was previously realized. The supposed influence of haltere extirpation on the wing-beat frequency has been shown by Faust (1952) to be incorrect for *Tipula* and *Calliphora*, but there remain unexplained 'Stimulationsorgane' effects (von Buddenbrock, 1919) in some of the nematocerous and brachycerous families (Brauns, 1939; Melin, 1941). Schneider's (1953) claim to have demonstrated a static sense mediated by the halteres of *Calliphora* has already been criticized, but he also states that they have a dynamic stabilizing function by virtue of a direct physical effect as opposed to a reflex mechanism, and this claim must first be examined in more detail. The evidence was of two types. Flies (*Calliphora*) were mounted so that they were free to rotate about a vertical axis formed by a short rod glued to the top of the thorax, and in the first set of experiments a statistical comparison was made between the rates of unstable rotation in intact and haltere-less flies performing 'anaesthetic flight'; those without halteres rotated faster. In the second set, radiant heat was used at a critical intensity to destroy the working of the basal sense organs without causing mechanical damage; the halteres of flies treated in this way still showed a stabilizing influence. The conclusion from these results, that there is a direct physical action of the halteres on the movements of the wings, relies on the assumption that haltere reflexes are absent in 'anaesthetic flight' and that the heat treatment had destroyed all sensory function. The reasoning is unconvincing in view of the extreme physical difficulties of a possible mechanism. Pringle (1948) showed that the mass of the halteres is about 0·04 % of the mass of the fly, so that a direct gyro-stabilizing effect is out of the question. Only by a form of servomechanical influence on the force-transmission mechanism in the basal wing articulation could a direct physical influence operate, and the anatomy of the fly does not suggest that such a mechanism is present. Schneider's conclusion cannot, therefore, be accepted on the available evidence.

Schneider (1953) has, however, good experimental evidence of a reflex stabilization in the yawing plane, and Faust (1952) extends this to pitch and, to some extent, to roll, with a detailed analysis of the nature of the compensatory wing movements. Faust's results were obtained by high-speed (3000 frames per sec.) cine-photography of a fly (*Calliphora*) mounted on the axis of a rotating shaft in the three definitive orientations. Some of his tracings from the photographs are reproduced in fig. 47.

The normal changes of the angle of twist in a stationary, tethered fly are shown in fig. 47 A—pronation during the downstroke and supination during the upstroke. There were often slight asymmetries in the instant of twisting at the two ends of the beat, and symmetrical differences in the angle of attack during the downstroke; these were thought to be active steering movements and variations in the lift (compare the fluctuation of the downstroke position in *Muscina*, fig. 6), but they could not be confused with the regulatory movements during rotations. The regulatory responses concerned only the downstroke; a full supination twist during the upstroke is a constant feature of all the photographs. The response to anti-clockwise rotation about the longitudinal axis (roll to the left at 2·1 rotations per sec.) is shown in fig. 47 B. At the end of the upstroke (*a*) the right wing pronates first; during the first part of the downstroke (*b*) the twist is symmetrical; during the rest of the stroke (*c*) the left wing has a higher angle of attack. The total effect must be to produce a clockwise torque tending to resist the rotation, and there is evidence that the magnitude of the effect depends on the velocity of rotation. Extirpation of the halteres reduces the response slightly, but it is still clearly present; blinding reduces it considerably and both mutilations remove it altogether. On rotations about the other two axes blinding has little effect. Fig. 47c shows the response to clockwise rotation about the vertical axis (yaw to the right) at 1·5 rotations per sec. The right wing pronates as usual at the start of the downstroke (d) and shows merely a slightly greater than normal angle of attack

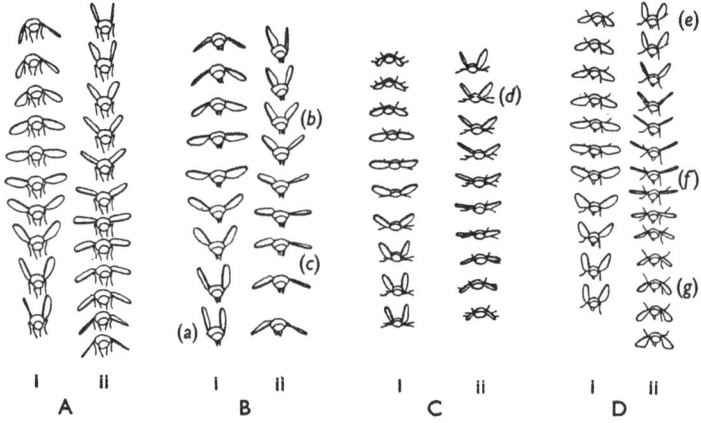

Fig. 47. Outline tracings from high-speed cinematographs of *Calliphora*, showing the wing twisting at different instants in the stroke. The fly is viewed head-on, and the anterior edge of the wing is drawn thicker than the posterior edge. Series i, upstroke; series ii, downstroke; read each series from top to bottom. A, normal tethered flight; B, rolling rotation to the left at 2·1 rotations per sec.; C, yawing rotation to the right at 1·5 rotations per sec.; D, forward pitching rotation at 2·5 rotations per sec. The letters indicate positions discussed in the text. (From Faust, 1952.)

throughout this stroke; the left wing fails to pronate and preserves the full supination twist during the whole of the downstroke. The total effect must be to generate a torque about the vertical axis resisting the yaw; the increased lift generated by the left wing may well be balanced by an increase in lift owing to the higher angle of attack of the right wing, so that no rolling moment results. Fig. 47D shows the response of a blinded fly to forward rotation about the transverse axis (forward pitch) at 2·5 rotations per sec. At the start of the downstroke (e) pronation is delayed in both wings; the angle of attack is greater than normal throughout the stroke (f); supination sometimes occurs before the end of the downstroke (g). The characteristic feature of the response is the supinated position at the start of the downstroke, which must produce a torque resisting the forward rotation; the effect of the early supination at the end of the downstroke is not clear and the phenomenon appeared to be variable. It seems (though Faust did not mention this) that there must be a considerable increase in the lift unless other parameters also vary. The experiments were not able to demonstrate any changes that may have occurred in the fore-and-aft position of the stroke plane, such as might be expected from the observations of Hollick (1940). Both the characteristic responses to yawing and pitching rotations were abolished by extirpation of the halteres.

These remarkable results would seem to establish beyond doubt that the halteres mediate a reflex response to rotations in the pitching and rolling, as well as the yawing planes, and Faust confirms that all these instabilities are shown by a free-flying, haltere-less fly. He agrees with Chadwick (1953) in disallowing Pringle's (1948) reasons for rejecting the possibility of a sensory indication of pitching rotations from the haltere sense organs; both authors correctly point out that these were based on the supposed occurrence of different frequencies of haltere vibration on the two sides in a flying insect. Schneider (1953) gives grounds for believing that such asynchrony never occurs during flight, when a powerful synchronizing force for haltere vibration is produced by the thoracic distortion due to activity of the indirect flight muscles, but only when the wings are stationary and the halteres are being vibrated solely by their own small, specific muscles. Sellke's (1936) observations of haltere asynchrony were made not on a muscid but on a non-flying *Tipula*, in which species Faust (1952) finds the rotation reflexes much less clearly demonstrable. If, as now appears proved, the halteres are always vibrated synchronously on the two sides during flight, it becomes necessary to re-examine their potentialities as special sense organs.

Pringle's (1948) analysis of the dynamics of haltere movement showed unequivocally that whereas the gyroscopic torques generated by yawing rotations are qualitatively distinct from anything present in the stationary insect because they oscillate at twice the frequency of haltere vibration (fig. 48), those generated during pitch and roll oscillate at the vibration frequency and are identical except that the torques in the two organs are in phase during pitch and in antiphase during roll. Stated differently, each haltere can give indications of rotations about only two axes, the vertical and a horizontal axis through its plane of vibration. Since the planes of vibration are different on the two sides, the intact insect with synchronous haltere vibration can, theoretically, resolve all rotations into their three components, but it cannot do so if one haltere is removed or if the synchronism is disturbed. This is a rigorous physical deduction and does not depend on observation.

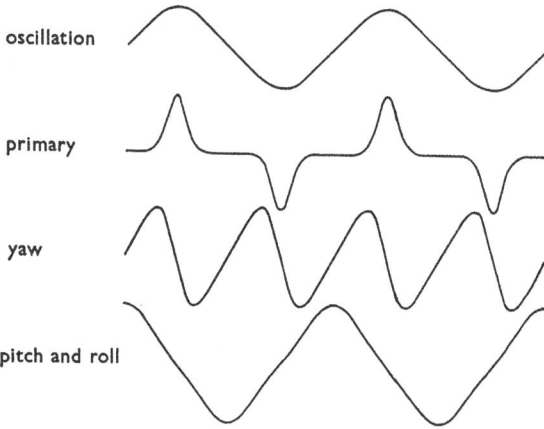

oscillation

primary

yaw

pitch and roll

Fig. 48. Calculated curves for the torques present at the base of a muscid haltere. The primary torque is in the stroke plane and is present when the fly is stationary. The two lower curves show the gyroscopic torques at right angles to the stroke plane during rotation of the whole fly. The relative amplitudes are of no significance. (From Pringle, 1948.)

It is known (Schelver, 1802; Pflugstaedt, 1912; Fraenkel, 1939) that removal of one haltere has either no effect or only a slight effect on flight; on the other hand, Faust (1952) found erratic changes in the pattern of wing movements when a blinded fly with one haltere was rotated in his apparatus. It is possible to resolve the problem in terms of a modified hypothesis, which is consistent with all the known facts, if we suppose that the individual haltere gives sensory indications of rotations about only two axes, but that the gyroscopic torques at haltere-vibration frequency from the two organs are summed in the ganglion in phase to release pitch-compensating movements and in antiphase to release roll-compensating movements. Intact flies should thus have in their halteres a mechanism for the discrimination of all three rotations, but in flies with only one haltere there must be confusion between the responses to roll and pitch; this explains the erratic compensatory wing movements found by Faust in such blinded flies. When supplemented by visual regulation the good haltere indications of yaw and the limited indications of pitch are, however, adequate to preserve stability in flight. It is to be expected that the flight of a fly with only one haltere would be more affected by darkness that that of the intact insect.

Faust (1952) finds it incomprehensible that the well-developed scapal groups of campaniform sensilla on the haltere base (R3, fig. 43) should have the function merely of regulating the amplitude of haltere vibration, as deduced by Pringle (1948). This shows a lack of appreciation of the importance of this parameter in a sense organ required to give accurate quantitative information about rotational velocities. The scapal and other sensilla which respond to the oscillatory torques always present in the vibrating haltere must provide a reference for central appreciation of the phasing of the gyroscopic torques; but, further than this, accurate relative sensitivity to yawing and pitching (or rolling) rotations demands a well-regulated amplitude of haltere vibration (ϕ_0), since the gyroscopic torques for pitch and roll depend on ϕ_0 and the yawing torques on ϕ_0^2. Greater elaboration of these sensilla groups is thus to be expected in the more skilful fliers. With the modification outlined above, Pringle's (1948) analysis of the working of the halteres therefore requires no further revision in the light of the work of Faust (1952) and Schneider (1953), except that the latter author has demonstrated in *Calliphora* a depressor as well as a levator haltere muscle.

THE HEAD AS AN ORGAN OF BALANCE. Faust (1952), in the first part of his paper, describes the results of experimental determinations of the ability of a variety of insects to fly in the dark; fully stable flight is found only in Diptera and Odonata. The influence of light was partially analysed by von Buddenbrock (1937) in terms of the 'Lichtrückenreaktion', an orientation depending on the greater light intensity falling on the upper part of the eyes of an insect. This reaction and the general role of the head as an organ of balance in Odonata have been studied in detail by Mittelstaedt (1950).

The head of a dragonfly is supported on a forwardly project-
ing spur of the prothorax by a median peg-joint and is held in
position by three pairs of muscles: levators, depressors and
rotators (fig. 49). There is, in addition, on each side wide of the
mid-line, a clamping knob on the prothorax which fits into a
cup on the head when this is drawn back by the levator muscles.
Unless this clamp is in operation the neck joint has a very free
movement, and since the head is a wide, heavy object rotations
of the body produce movement at the joint which is detected by

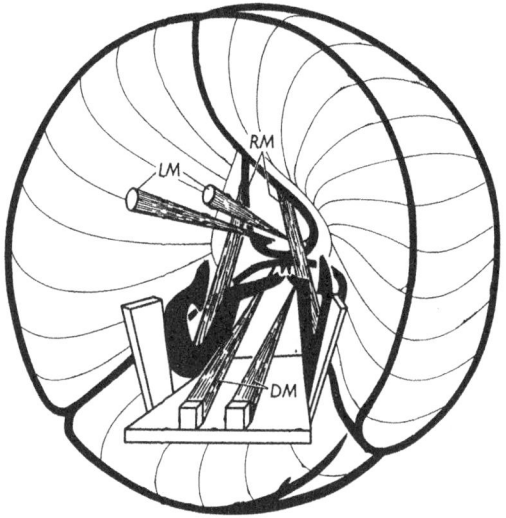

Fig. 49. Diagram to illustrate the head suspension of a dragonfly; *DM*, depressor
muscles; *LM*, levator muscles; *RM*, rotator muscles. (From a model of *Anax
imperator*; Mittelstaedt, 1950.)

four groups of prothoracic hair plates which are deflected by
contact with the head. Mittelstaedt showed experimentally
that twisting the head (by a magnetic pull on a small iron rod
glued to the frons) of a tethered, flying insect (*Anax imperator*)
induces compensatory wing movements to bring the body back
into line with the head. He also demonstrated that this response
is still present when the wings are not vibrating and was thus
able to show that the effective movement is a differential twist
of all four wings (fig. 50); the habit of the insect of perching in
exposed positions with the wings extended probably makes this
a useful balancing reaction.

Electrophysiological investigation of similar hair plates on the legs of *Periplaneta* (Pringle, 1938 *b*) showed that the integrated excitation of the many endings could give an indication of joint position. The two pairs of prothoracic hair plates of *Anax imperator* are so arranged that all are partially excited in the resting position of the head; during left-handed (anticlockwise from behind) rotation of the head the left 'spur plate' and the right 'neck plate' are more strongly stimulated, and the other two less strongly. By selective section of their sensory nerves, Mittelstaedt was able to show that the crossed pair of hair plates acts synergically during rolling movements. There is thus present a dynamic sense for rolling rotations.

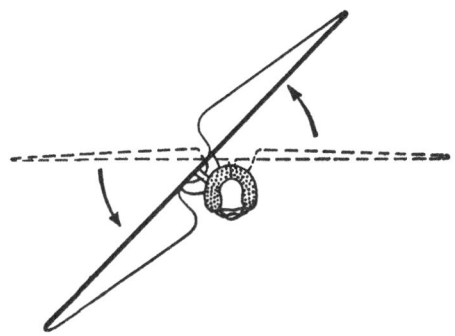

Fig. 50. Compensatory wing-twisting in *Anax* when the body is displaced; the head remains stationary, and the wing-twisting (hind wings only drawn) is such as to bring the body back to the normal position (broken lines). (From Mittelstaedt, 1950.)

The 'Lichtrückenreaktion' operates in the first instance to produce rotation of the head by means of the neck muscles. This head displacement then releases the wing responses. The intact insect flying in the open thus orientates its head in relation to the light gradient and its body in relation to its head; in the dark the inertia of the head is evidently sufficient to give it a dynamic sense which preserves stability.

Prothoracic hair plates are found in other insects besides the Odonata, including *Calliphora* (Lowne, 1890) and *Mantis* (Pringle, 1938 *b*), and it is probable that regulation in roll is often achieved with their assistance; Faust's (1952) demonstration of a visual stabilization in roll in *Calliphora* can be explained in the same way as that of *Anax*. Mittelstaedt (1950) does not discuss the possibility that the same two pairs of hair plates in

Anax might also mediate a stabilization in pitch or yaw by a different central interpretation of their relative excitation; there is no experimental evidence for such a reflex mechanism, but it is difficult otherwise to find functional explanation of the presence of four hair plates rather than merely two which would be adequate for regulation in a single plane.

Optomotor responses in the yawing and rolling planes certainly occur during flight as well as during crawling (Schneider (1956) for *Calliphora*), but it seems likely that in many insects this is a direct effect of the visual stimuli on the thoracic motor pattern and not an indirect effect through head movement; Wolf and Zerrahn-Wolf (1936) found that a bee (*Apis mellifera*), held so that only the head was free to move, reacted to a moving pattern of vertical stripes with antennal pointing but not with head movement. On the other hand, Autrum and Stöcker (1952) report nystagmus during rotation of the visual field in *Vespa*. The reflex aspect of these optomotor reactions requires further analysis.

INNATE PATTERNS OF CO-ORDINATION. The constancy of the pattern of muscular contractions found in insect flight suggests that there may be an innate organization of central connexions in the nervous system responsible for the basic rhythm of activity and modified by reflex actions. This question is discussed by Weis-Fogh (1956 b) in relation to the flight of *Schistocerca*. He finds that, while the duration and speed of contractions in isolated muscles are dependent on temperature, the frequency of wing beats and many other flight parameters in the intact insect are independent, and he therefore concludes that the rhythm must be determined reflexly by reference to some temperature-independent feature of the flight system (such as the duration of the upstroke, which is affected only by wing inertia). The possibility cannot be excluded that the relative phasing of excitation to the different muscles may be due to an innate central organization, but here again a reflex explanation is more probable. It is difficult to see how further evidence on this subject can be obtained until we have a much more complete knowledge of the neuro-anatomy and neuro-physiology of the flight system.

SUMMARY OF REGULATORY SYSTEMS IN INSECT FLIGHT. Besides the work already reviewed in this chapter, there are a number of papers which deal briefly with some aspects of the regulation of flight. That by Stellwaag (1916) is much quoted. Stellwaag observed in a variety of species that the abdomen might be displaced and the legs extended to a different extent on the two sides during rotation of a suspended flying insect; this was confirmed by Hollick (1940) and Faust (1952) (fig. 47). Of the wing-beat parameters Stellwaag emphasizes particularly the changes which can occur in the stroke plane, and he suggests that in the bee, *Apis mellifera*, this may be used both for lateral control and for the adjustment of forward motion (fig. 51). The stroke-plane angle does not vary appreciably in *Schistocerca* (Weis-Fogh, 1956a); slow or hovering flight is achieved by many insects (*Macroglossum* and Coleoptera) at least partly by a pronounced backward tilt of the whole body. The Apoidea may be exceptional in this respect, but other methods of control in roll and yaw than that of fig. 51 A are possible through the action of the direct muscles.

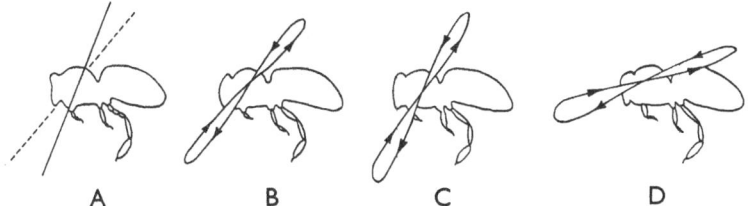

A   B   C   D

Fig. 51. Variation of the stroke plane in *Apis mellifera*, according to Stellwaag (1916). A, differential change of stroke plane for turning movements; B, hovering; C, flying forwards; D, flying backwards.

Flight regulation in dragonflies after extensive asymmetrical wing-cutting is described by Kamada and Kinosita (1947). The ability to fly straight after release improved on repetition but disappeared again after 5–10 min. if the insect was restrained with or without accompanying wing movement. This looks like a true learned compensation to visual stimuli, but the evidence is insufficient for any definite conclusion.

Demoll's (1918) experiments with beetles (*Melolontha*) were mentioned earlier. The doubling of the speed of flight on removal of 45 % of the elytra suggests a reflex regulating wing movements in relation to the total aerodynamic force on elytra and wings; when the area of the wings was reduced by 25 %, the reduction in forward speed to half the normal could again be brought about by a lift-regulating reflex, since at the (observed) greater body angle the thrust component of the aerodynamic force would be reduced.

The important regulatory systems are summarized in fig. 52, in which the features regulated are shown by heavy circles, diamonds or rectangles and the regulatory influences by light squares. The diagram is largely self-explanatory, but it should be noted that frequency is shown as an interrupted diamond, since it is maintained that this parameter is not an important feature of the nervous regulation of flight; it is set for any given

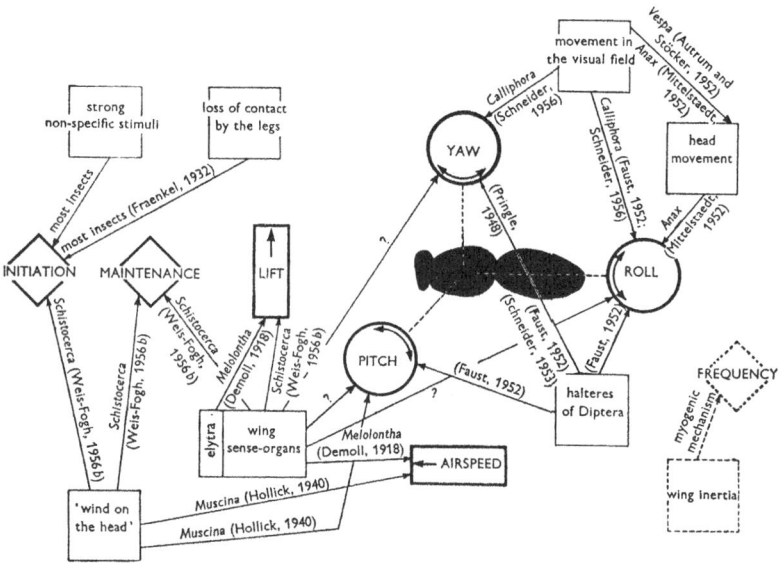

Fig. 52. Diagrammatic summary of reflex regulatory mechanisms in insect flight.

insect by physical dimensions and its variation in the individual is largely a corollary of the flight-maintenance system in neurogenic species and of the muscle physiology in myogenic species. The regulation of angular movements (roll, pitch, yaw) from wing sense-organs is shown with a question mark, since no conclusive experimental proof is available; it is rendered highly probable by the demonstration of such regulation from the homologous halteres of Diptera, and on the general grounds that many insects not possessing halteres have a body shape which is unlikely to confer inherent aerodynamic stability.

REFERENCES

ABBOTT, B. C., BIGLAND, B. and RITCHIE, J. M. (1952). The physiological cost of negative work. *J. Physiol.* **117**, 380–90.

ALBRECHT, F. O. (1953). *The Anatomy of the Migratory Locust.* London.

ALEXANDROWICZ, J. S. (1951). Muscle receptor organs in the abdomen of *Homarus vulgaris* and *Palinurus vulgaris. Quart. J. micr. Sci.* **92**, 163–99.

ATTILA, U. (1947). Betrachtung des Flügelschlags bei Insekten an Hand eines physikalischen Modells. *Acta ent. fenn.* **5**, 1–9.

AUTRUM, H. and SCHNEIDER, W. (1948). Vergleichende Untersuchungen über den Erschütterungssinn der Insekten. *Z. vergl. Physiol.* **31**, 77–88.

AUTRUM, H. and STÖCKER, M. (1952). Über optische Verschmelzungsfrequenzen und stroboskopisches Sehen bei Insekten. *Biol. Zbl.* **71**, 129–52.

BALDWIN, E. and NEEDHAM, D. M. (1934). The phosphate distribution in resting fly muscle. *J. Physiol.* **80**, 221–37.

BARRON, E. S. G. and TAHMISIAN, T. N. (1948). The metabolism of cockroach muscle. *J. cell. comp. Physiol.* **32**, 57–76.

BEALL, G. (1948). The fat content of a butterfly, *Danaus plexippus* Linn., as affected by migration. *Ecology*, **29**, 80–94.

BEHRENDT, R. (1940). Untersuchungen über die Wirkungen erblichen und nichterblichen Fehlens bzw. Nichtgebrauchs der Flügel auf die Flugmuskulatur von *Drosophila melanogaster. Z. wiss. Zool.* **152**, 129–58.

BEUTLER, R. (1937). Über den Blutzucker der Biene. *Z. vergl. Physiol.* **24**, 71–115.

BINET, A. (1894). Contributions à l'étude du systèmes nerveux sous-intestinal des insectes. *J. Anat., Paris*, **30**, 449–580.

BOETTIGER, E. G. (1951). Stimulation of the flight muscles of the fly. *Anat. Rec.* **111**, 443.

BOETTIGER, E. G. (1955). Triggering the contraction process in insect fibrillar muscle. *J. cell. comp. Physiol.* **46**, 370–1.

BOETTIGER, E. G. (1957a). The machinery of insect flight. In *Recent Advances in Invertebrate Physiology* (ED. Scheer, B. T.), pp. 117–42. University of Oregon Publications.

BOETTIGER, E. G. (1957b). Triggering of the contractile process in insect fibrillar muscle. In *Physiological Triggers* (ED. Bullock, T. H.), pp. 103–16. American Physiological Society.

BOETTIGER, E. G. and FURSHPAN, E. (1950). Observations on the flight motor of Diptera. *Biol. Bull., Woods Hole*, **99**, 346.

BOETTIGER, E. G. and FURSHPAN, E. (1951). Observations on the flight motor of Diptera. *Fed. Proc.* **10**, 17.

BOETTIGER, E. G. and FURSHPAN, E. (1952). The mechanics of flight movements in Diptera. *Biol. Bull., Woods Hole*, **102**, 200–211.

BOETTIGER, E. G. and FURSHPAN, E. (1954). Mechanical properties of insect flight muscle. *J. cell. comp. Physiol.* **44**, 340.

BOETTIGER, E. G. and McCANN, F. (1953). Single fibre action potentials in insect fibrillar muscle. *Fed. Proc.* **12**, 17.

BONHAG, P. F. (1949). The thoracic mechanism of the adult horsefly (Diptera: Tabanidae). *Mem. Cornell agric. Exp. Sta.*, no. 285.

BRAUNS, A. (1939). Morphologische und physiologische Untersuchungen zum Halterenproblem, unter besonderer Berücksichtigung brachytyper Arten. *Zool. Jb.*, Allg. Zool. Physiol., **59**, 245–390.

BUCHTHAL, F. and WEIS-FOGH, T. (1956). Contribution of the sarcolemma to the force exerted by resting muscle of insects. *Acta physiol. Scand.* **35**, 345–64.

BUCHTHAL, F., WEIS-FOGH, T. and ROSENFALCK, P. (1957). Twitch contractions of isolated flight muscle of locusts. *Acta physiol. Scand.* **39**, 246–276.

VON BUDDENBROCK, W. (1919). Die vermutliche lösung der Halterenfrage. *Pflüg. Arch. ges. Physiol.* **175**, 125–64.

VON BUDDENBROCK, W. (1937). *Grundniss der vergleichenden Physiologie.* Berlin.

CARBONELL, C. S. (1947). The thoracic muscles of the cockroach. *Smithson. misc. Coll.* **107** (2), 1–23.

CARPENTIER, F. (1923). Musculature et squelette chitineus. *Mem. Acad. roy. Belg.*, Cl. Sci., **7**, 1–56.

CHABRIER, J. (1822). *Essai sur le vol des insectes.* Paris.

CHADWICK, L. E. (1940). The wing motion of the dragonfly. *Bull. Brooklyn ent. Soc.* **35**, 109–12.

CHADWICK, L. E. (1947). The respiratory quotient of *Drosophila* in flight. *Biol. Bull., Woods Hole*, **93**, 229–39.

CHADWICK, L. E. (1951). Stroke amplitude as a function of air density in the flight of *Drosophila*. *Biol. Bull., Woods Hole*, **100**, 15–27.

CHADWICK, L. E. (1953). The motion of the wings. Aerodynamics and flight metabolism. The flight muscles and their control. In Roeder, K. D., *Insect Physiology*. New York.

CHADWICK, L. E. and GILMOUR, D. (1940). Respiration during flight in *Drosophila repleta*: the oxygen consumption considered in relation to the wing-rate. *Physiol. Zoöl.* **13**, 398–410.

CHADWICK, L. E. and WILLIAMS, C. M. (1949). The effects of atmospheric pressure and composition on the flight of *Drosophila*. *Biol. Bull., Woods Hole*, **97**, 115–137.

CHAO, H.-F. (1953). The external morphology of the dragonfly *Onchogramphus ardens* Needham. *Smithson. misc. Coll.* **122**, no. 6.

CHAPMAN, G. B. (1954). Electron microscopy of ultra-thin sections of insect flight muscle. *J. Morph.* **95**, 237–51.

CIACCIO, G. (1940). Richerche sull azione degli agenti chemici sulle fibre dei muscoli delle ali di alcuni coleopteri. *Z. Zellforsch.* **30**, 567–594.

CIACCIO, G. V. (1887). *Mem. R. Accad. Bologna, Sci. fis.* (iv S), **8**, 525, (quoted by Tiegs, 1955).

CLARK, H. W. (1940). The adult musculature of the Anisopterous dragonfly thorax (Odonata, Anisoptera). *J. Morph.* **67**, 523–565.

CLELAND, K. W. and SLATER, E. C. (1953). The sarcosomes of heart muscle. *Quart. J. micr. Sci.* **94**, 329–346.

COMSTOCK, J. H. (1918). *The Wings of Insects.* Ithaca.

COMSTOCK, J. H. and NEEDHAM, J. G. (1898). The wings of insects. *Amer. Nat.* **32**, 42, 81, 231, 335, 413, 561, 768, 903.

COMSTOCK, J. H. and NEEDHAM, J. G. (1899). The wings of insects. *Amer. Nat.* **33**, 117, 573, 845.

CRAMPTON, G. C. (1927). The thoracic sclerites and wing bases of the roach *Periplaneta americana* and the basal structures of the wings of insects. *Psyche, Lond.*, **34**, 59–72.

CREMER, E. (1934). Anatomische, reizphysiologische und histologische Untersuchungen an der imaginalen und larvalen Flugmuskulatur der Odonaten. *Zool. Jb.*, Allg. Zool. Physiol., **54**, 191–223.

DANZER, A. (1956). Der Flugapparat der Dipteren als Resonanzsystem. *Z. vergl. Physiol.* **38**, 259–83.

DAVIS, R. A. and FRAENKEL, G. (1940). The oxygen consumption of flies during flight. *J. exp. Biol.* **17**, 402–7.

DEMOLL, R. (1918). *Der Flug der Insekten und der Vögel*. Jena: Gustav Fischer.

DEMOLL, R. (1919). Der Flug der Insekten. *Naturwissenschaften*, **7**, 480–1.

DERHAM, W. (1713). *Physico-Theology*. London: Boyle Lecture for 1711.

DIAKONOFF, A. (1936). Contributions to the knowledge of the fly reflexes and the static sense in *Periplaneta americana*. *Arch. néerl. Physiol.* **21**, 104–29.

DOTTERWEICH, K. (1928). Beiträge zur Nervenphysiologie der Insekten. *Zool. Jb.*, Allg. Zool. Physiol., **44**, 399–425.

EDWARDS, G. A. and RUSKA, H. (1953). The function and metabolism of certain insect muscles in relation to their structure. *Quart. J. micr. Sci.* **96**, 151–9.

EDWARDS, G. A., SANTOS, P. de S., SANTOS, H. L. de S., and SAWAYA, P. (1954*a*). Electron microscope studies of insect muscle. I. Flight and coxal muscle of *Hydrophilus piceus*. *Ann. ent. Soc. Amer.* **47**, 343–54.

EDWARDS, G. A., SANTOS, P. de S., SANTOS, H. L. de S., and SAWAYA, P. (1954*b*). Electron microscope studies of insect muscle. II. Flight and leg muscles of *Belostoma* and *Periplaneta*. *Ann. ent. Soc. Amer.* **47**, 459–67.

EDWARDS, G. A., SANTOS, P. de S., SANTOS, H. L. de S., and SAWAYA, P. (1954*c*). Electron microscope studies of insect muscle. III. Variations in ultra-structure. *Bol. Fac. Filos. Ciênc. S. Paulo*, **19**, 391–405.

EGGERS, F. (1928). Die Stiftführenden Sinnesorgane. *Zool. Baust.* **2**, no. 1.

EHNBOM, K. (1948). Studies on the central and sympathetic nervous system and some sense organs in neuropterous insects. *Opusc. ent.* (Suppl.) **8**.

ELTRINGHAM, H. (1925). On a new organ in certain Lepidoptera. *Trans. R. ent. Soc. Lond.* **73**, 7–9.

ERHARDT, E. (1916). Zur kenntnis der Innervierung und der Sinnesorgane der Flügel von Insekten. *Zool. Jb.*, Anat. u. Ontog., **39**, 293–334.

EWER, D. W. (1953). The anatomy of the nervous system of the tree locust, *Acanthacris ruficornis* (Fab.). I. Adult metathorax. *Ann. Natal Mus.* **12**, 367–81.

EWER, D. W. (1954*a*). The anatomy of the nervous system of the tree locust, *Acanthacris ruficornis* (Fab.). II. The adult mesothorax. *J. ent. Soc. S. Afr.* **17**, 27–37.

EWER, D. W. (1954*b*). A note on the comparative anatomy of the ptero-thorax of macropterous and brachypterous forms of the grasshopper, *Zonocerus elegans* Thunb. *J. ent. Soc. S. Afr.* **17**, 237–40.

EWER, D. W. and RIPLEY, S. H. (1953). On certain properties of the flight muscles of the Orthoptera. *J. exp. Biol.* **30**, 170–7.

FARREN, W. S. (1936). The reaction on a wing whose angle of incidence is changing rapidly. *Rep. aero. Res. Comm., Lond.*, no. 1648, H.M. Stationery Office.

FAUST, R. (1952). Untersuchungen zum Halterenproblem. *Zool. Jb.*, Allg. Zool. Physiol., **63**, 325–66.

FINLAYSON, L. H. and LOWENSTEIN, O. (1955). A proprioceptor in the body musculature of Lepidoptera. *Nature, Lond.*, **176**, 1031.

FORBES, W. T. M. (1943). The origin of wings and venational types in insects. *Amer. Midl. Nat.* **29**, 381–405.

FRAENKEL, G. (1932). Untersuchungen über die Koordination von Reflexen und automatisch-nervösen Rhythmen bei Insekten. I. Die Flugreflexe der Insekten und ihre Koordination. *Z. vergl. Physiol.* **16**, 371–93.

FRAENKEL, G. (1939). The function of the halteres of flies (Diptera). *Proc. zool. Soc. Lond.* A, **109**, 69–78.

FULLER, C. (1925). The thorax and abdomen of winged termites with special reference to the sclerites and muscles of the thorax. *Ent. Mem. S. Afr. Dep. Agric.* **2**, 49–78.

FULTON, R. A. and ROMNEY, V. E. (1940). The chloroform-soluble com-ponents of beet leafhoppers as an indication of the distance they move in the spring. *J. agric. Res.* **61**, 737–43.

GASSER, H. S. and HILL, A. V. (1924). The dynamics of muscular contrac-tion. *J. Physiol.* **96**, 398–437.

GILMOUR, D. (1953). Localization of the magnesium-activated apyrase of insect muscle in the sarcosomes. *Austr. J. Sci. Res.* B, **6**, 586–90.

GILMOUR, D. and CALABY, J. H. (1953*a*). Physical and enzymatic properties of actomyosins from the femoral and thoracic muscles of an insect. *Enzymologia*, **16**, 23–33.

GILMOUR, D. and CALABY, J. H. (1953*b*). Myokinase and pyrophosphatase of insect muscle. *Enzymologia*, **16**, 34–40.

GLAUERT, H. (1935). Airplane propellers. In Durand, W. F. (ED.) *Aerodynamic theory*, **4**, 169–360.

GOODALL, M. C. (1956). Auto-oscillations in extracted muscle fibre systems. *Nature*, **177**, 1238–9.

GRAY, J. and HANCOCK, G. J. (1955). The propulsion of sea-urchin spermatozoa. *J. exp. Biol.* **32**, 802–14.

HAGIWARA, S. (1953). Neuro-muscular transmission in insects. *Jap. J. Physiol.* **3**, 284–296.

HAGIWARA, S. and WATANABE, A. (1954). Action potential of insect muscle examined with intracellular electrodes. *Jap. J. Physiol.* **4**, 65–78.

HANSON, J. (1952). Changes in the cross-striation of myofibrils during con-traction induced by adenosine triphosphate. *Nature, Lond.*, **169**, 530–3.

HANSON, J. (1956*a*). Elongation of cross-striated myofibrils. *Biochem. biophys. Acta*, **20**, 289–92.

HANSON, J. (1956b). Studies on the cross-striation of the indirect flight myofibrils of the blowfly *Calliphora*. *J. biophys. biochem. Cytol.* **2**, 691–710.

HANSON, J. and HUXLEY, H. E. (1955). The structural basis of contraction in striated muscle. *Symp. Soc. exp. Biol.* **9**, 228–64.

HASKEN, W. (1939). Der Thorax von *Panorpa communis* L. *Zool. Jb.*, Anat. Physiol., **65**, 295–338.

HEIDERMANNS, C. (1931). Reizphysiologische Untersuchungen an der Flugmuskulatur von *Aeschna coerulea*. *Zool. Jb.*, Allg. Zool. Physiol., **50**, 1–31.

HERTWECK, H. (1931). Anatomie und Variabilität des Nervensystems und der Sinnesorgane von *Drosophila melanogaster* (Meigen). *Z. wiss. Zool.* **139**, 559–663.

HILL, A. V. (1951). The influence of temperature on the tension developed in an isometric twitch. *Proc. Roy. Soc.* B, **138**, 349–54.

HOCKING, B. (1953). The intrinsic range and speed of insect flight. *Trans. R. ent. Soc. Lond.* **104**, 223–345.

HODGE, A. J. (1955). Studies on the structure of muscle. III. Phase contrast and electron microscopy of dipteran flight muscle. *J. biophys. biochem. Cytol.* **1**, 361–80.

HODGE, A. J. (1956). The fine structure of striated muscle. A comparison of insect flight muscle with vertebrate and invertebrate skeletal muscle. *J. biophys. biochem. Cytol.* **2**, (Suppl.) 131–42.

HOLLICK, F. S. J. (1940). The flight of the dipterous fly *Muscina stabulans* Fallén. *Phil. Trans.* B, **230**, 357–90.

HOLMGREN, E. (1910). Untersuchungen über die morphologische nachweisbaren stofflichen Umsetzungen der quergestreiften Muskelfasern. *Arch. mikr. Anat.* **75**, 240–336.

HOLST, E. v. and KÜCHEMANN, D. (1941). Biologische und aerodynamische Probleme des Tierfluges. *Naturwissenschaften*, **29**, 348–62.

HOLST, E. v. and KÜCHEMANN, D. (1942). Biological and aerodynamic problems in animal flight. *J. R. aero. Soc.* **46**, 39–56.

HORRIDGE, A. (1956). The flight of very small insects. *Nature, Lond.*, **178**, 1334–5.

HOYLE, G. (1955a). The anatomy and innervation of locust skeletal muscle. *Proc. Roy. Soc.* B, **143**, 281–92.

HOYLE, G. (1955b). Neuromuscular mechanisms of a locust skeletal muscle. *Proc. Roy. Soc.* B, **143**, 343–67.

HUGHES, G. M. (1952). Abdominal mechanoreceptors in *Dytiscus* and *Locusta*. *Nature, Lond.*, **170**, 531–2.

HUMPHREY, G. F. (1949). Invertebrate glycolysis. *J. cell. comp. Physiol.* **34**, 323–5.

HUMPHREY, G. F. and SIGGINS, L. (1949). Glycolysis in the wing muscle of the grasshopper, *Locusta migratoria*. *Aust. J. exp. Biol. med. Sci.* **27**, 353–9.

JACKSON, D. J. (1933). Observations on the flight muscles of *Sitona* weevils. *Ann. appl. Biol.* **20**, 731–70.

JACKSON, D. J. (1952). Observations on the capacity for flight of water beetles. *Proc. R. ent. Soc. Lond.* A, **27**, 57–70.

JANET, C. (1899). Sur le mécanisme du vol chez les Insectes. *C. R. Acad. Sci., Paris*, **128**, 249–53.

Jasper, H. H. and Pezard, A. (1934). Relation entre la rapidité d'un muscle strié et sa structure histologique. *C. R. Acad. Sci., Paris*, **198**, 499–501.

Jawlowsky, H. (1936). Über den Gehirnbau der Käfer. *Z. Morph. Ökol. Tiere*, **32**, 67–91.

Jensen, Martin (1956). Biology and physics of locust flight. III. The aerodynamics of locust flight. *Phil. Trans.* B, **239**, 511–52.

Jongbloed, J. and Wiersma, C. A. G. (1935). Der Stoffwechsel der Honigbiene während des Fliegens. *Z. vergl. Physiol.* **21**, 519–33.

Jordan, H. E. (1920). Studies on striped muscle structure. VI. The comparative histology of the leg and wing muscle structure of the wasp, with special reference to the phenomenon of stripe reversal during contraction and to the genetic relation between contraction bands and intercalated discs. *Amer. J. Anat.* **27**, 1–66.

Jordan, H. E. (1933). The structural changes in striped muscle during contraction. *Physiol. Rev.* **13**, 301–24.

Kamada, T. and Kinosita, H. (1947). Regulation of flight in dragonflies (in Japanese, English summary). *Seiro-Seitai (Physiology and Ecology)*, **1**, 147–59.

Keilich, J. (1918). Beiträge zur Kenntnis der Insektenmuskeln. *Zool. Jb.*, Anat. u. Ontog., **40**, 515–36.

Keilin, D. (1925). On cytochrome, a respiratory pigment common to animals, yeast and higher plants. *Proc. Roy. Soc.* B, **98**, 312–39.

Korschelt, E. (1923). *Bearbeitung Einheimischer Tiere. Der Gelbrand Dytiscus marginalis L.* Leipzig.

Korschelt, E. (1938). Cuticularsehne und Bindegewebssehne. Eine vergleichend morphologische-histologische Betrachtung. *Z. wiss. Zool.* **150**, 494–526.

Kraemer, K. (1932). Reizphysiologische Untersuchungen an Coleopteren-Muskulatur. *Zool. Jb.*, Allg. Zool. Physiol., **51**, 321–96.

Krogh, A. and Weis-Fogh, T. (1951). The respiratory exchange of the desert locust (*Schistocerca gregaria*) before, during and after flight. *J. exp. Biol.* **28**, 344–57.

Krogh, A. and Zeuthen, E. (1941). The mechanism of flight preparation in some insects. *J. exp. Biol.* **18**, 1–10.

Lehr, R. (1914). Die sinnesorgane der beiden Flügelpaare von *Dytiscus marginalis*. *Z. wiss. Zool.* **110**, 87–150.

Lewis, S. E. and Slater, E. C. (1953). Oxidative phosphorylation in insect sarcosomes. *Biochem. J.* **55**, xxvii.

Lissmann, H. W. (1946). The neurological basis of the locomotory rhythm in the spinal dogfish. II. The effect of de-afferentation. *J. exp. Biol.* **23**, 162–76.

Lorand, L. and Moos, C. (1956). Auto-oscillations in extracted muscle fibre systems. *Nature, Lond.*, **117**, 1239.

Lowne, B. T. (1890). *The Blow-fly.* London.

McEnroe, W. (1953). Tension-length curve of insect fibrillar muscle. *Fed. Proc.* **11**, 104.

McShan, W. H., Kramer, S. and Schlegel, V. (1954). Oxidative enzymes in the thoracic muscles of the woodroach, *Leucophaea maderae. Biol. Bull.*, Woods Hole, **106**, 341–52.

MAGNAN, A. (1934). *Le Vol des Insectes*. Paris.

MAGNAN, A. and SAINTE-LAGUË, A. (1933). Le vol au point fixe. *Actualités sci. industr.* **60**, 1–31.

MAKI, T. (1936). Studies of the skeletal structure, musculature and nervous system of the Alder fly, *Chauliodes formosanus* Peterson. *Mem. Fac. Sci. Agric. Taihoku*, **16**, 117–243.

MAKI, T. (1938). Studies on the thoracic musculature of insects. *Mem. Fac. Sci. Agric. Taihoku*, **24**, 1–343.

MANGOLD, E. (1905). Untersuchungen über die Endigung der Nerven in den quergestreiften Muskeln der Arthropoden. *Z. allg. Physiol.* **5**, 135–205.

MARCUS, H. (1920). Uber die Struktur und Entwicklung der quergestreifte Muskelfasern, besonders bei Flugmuskeln der Libellen. *Anat. Anz.* **52**, 410–16.

MAREY, É. J. (1868*a*). Determination experimentale du mouvement des ailes des insectes pendant le vol. *C. R. Acad. Sci. Paris*, **67**, 1341–5.

MAREY, É. J. (1868*b*). Memoire sur le vol des insectes et des oiseaux. *Ann. Sci. nat. Zool.* **12**, 49–150.

MATHESON, R. and CROSBY, C. R. (1912). Aquatic hymenoptera in America. *Ann. ent. Soc., Amer.* **5**, 65–71.

MATTHEWS, A. (1872). *A Monograph of the Trichopterygia*. London.

MELIN, D. (1941). Contributions to the knowledge of the flight of insects. *Uppsala Univ. Årsskr.* no. 4, 1–247.

MIHALYI, F. (1935–6). Untersuchungen über Anatomie und Mechanik der Flugorgane an der Stubenfliege. *Arb. ung. biol. ForschInst.* **8**, 106–119.

MILLER, A. (1950). The internal anatomy and histology of the imago of *Drosophila melanogaster*. In Demerec, M., *Biology of Drosophila*. New York.

MITTELSTAEDT, H. (1950). Physiologie des Gleichgewichtsinnes bei fliegenden Libellen. *Z. vergl. Physiol.* **32**, 422–63.

MORISON, G. D. (1928). The muscles of the adult honey-bee. *Quart. J. micr. Sci.* **71**, 395–463, 563–651.

NUESCH, H. (1953). The morphology of the thorax of *Telea polyphemus* (Lepidoptera). I. Skeleton and muscles. *J. Morph.* **93**, 589–609.

NUESCH, H. (1954). Segmentierung und Muskelinnervation bei *Telea polyphemus* (Lep.). *Rev. suisse Zool.* **61**, 420–8.

OSBORNE, M. F. M. (1951). Aerodynamics of flapping flight with application to insects. *J. exp. Biol.* **28**, 221–45.

OSSIANNILSSON, F. (1949). Insect drummers. *Opusc. ent.* (Suppl.), **10**.

PÉREZ-GONZÁLEZ, M. D. and EDWARDS, G. A. (1954). Metabolic differences among several specialized insect muscles. *Bol. Fac. Filos. Ciênc. S. Paulo*, **19**, 373–89.

PFLUGSTAEDT, H. (1912). Die Halteren der Dipteren. *Z. wiss. Zool.* **100**, 1–59.

PHILPOTT, D. E. and SZENT-GYÖRGYI, A. (1955). Observations on the electron microscope structure of insect muscle. *Biochem. biophys. Acta*, **18**, 177–82.

PRINGLE, J. W. S. (1938*a*). Proprioception in insects. I. A new type of mechanical receptor from the palps of the cockroach. II. The action of the campaniform sensilla on the legs. *J. exp. Biol.* **15**, 101–13, 114–31.

PRINGLE, J. W. S. (1938*b*). Proprioception in insects. III. The function of the hair sensilla at the joints. *J. exp. Biol.* **15**, 467–73.

125

PRINGLE, J. W. S. (1939). The motor mechanism of the insect leg. *J. exp. Biol.* **16**, 220–31.

PRINGLE, J. W. S. (1940). The reflex mechanism of the insect leg. *J. exp. Biol.* **17**, 8–17.

PRINGLE, J. W. S. (1948). The gyroscopic mechanism of the halteres of Diptera. *Phil. Trans.* B, **233**, 347–84.

PRINGLE, J. W. S. (1949). The excitation and contraction of the flight muscles of insects. *J. Physiol.* **108**, 226–32.

PRINGLE, J. W. S. (1950). The flight of insects. *School Sci. Rev.* **31**, 364–9.

PRINGLE, J. W. S. (1954*a*). The mechanism of the myogenic rhythm of certain insect striated muscles. *J. Physiol.* **124**, 269–91.

PRINGLE, J. W. S. (1954*b*). A physiological analysis of cicada song. *J. exp. Biol.* **31**, 525–60.

PRINGLE, J. W. S. (1956). The physiology of insect song. *Acta physiol. pharm. néerl.* **5**, 88–97.

PRINGLE, J. W. S. (1957). Myogenic rhythms. In *Recent Advances in Invertebrate Physiology* (ED. Scheer, B. T.) University of Oregon Publications pp. 99–115.

PUMPHREY, R. J. and RAWDON-SMITH, A. F. (1936). Hearing in insects: the nature of the response of certain receptors to auditory stimuli. *Proc. Roy. Soc.* B, **121**, 18–27.

RAMSEY, R. W. and STREET, S. F. (1940). The isometric length-tension diagram of isolated skeletal muscle fibres of the frog. *J. cell. comp. Physiol.* **15**, 11–34.

REES, K. R. (1954). Aerobic metabolism of the muscle of *Locusta migratoria*. *Biochem. J.* **58**, 196–202.

RETZIUS, G. (1890). Muskelfibrille und sarcoplasm. *Biol. Untersuch.* **1**, 51–88.

RITCHIE, J. M. and WILKIE, D. R. (1955). The effect of previous stimulation on the active state of muscle. *J. Physiol.* **130**, 488–96.

RITTER, W. (1912). The flying apparatus of the blow-fly. *Smithson. misc. Coll.* **56**.

ROCH, F. (1922). Beitrag zur Physiologie der Flugmuskulatur der Insekten. *Biol. Zbl.* **42**, 359–64.

ROEDER, K. D. (1951). Movements of the thorax and potential changes in the thoracic muscles of insects during flight. *Biol. Bull., Woods Hole*, **100**, 95–106.

ROEDER, K. D. and WEIANT, E. A. (1950). The electrical and mechanical events of neuro-muscular transmission in the cockroach, *Periplaneta americana* (L.). *J. exp. Biol.* **27**, 1–13.

RÜSCHKAMP, P. F. (1927). Der Flugapparat der Käfer, Vorbedingung, Ursache und Verlauf seiner Ruckbildung. *Zoologica*, **28**, Hft. 75, 1–88.

RUSSELL, H. M. (1912). The greenhouse thrips. *Circ. U.S. Bur. Ent.*, no. 151.

SAKTOR, B. (1953*a*). Investigations on the mitochondria of the house-fly, *Musca domestica* L. I. Adenosinetriphosphates. *J. gen. Physiol.* **36**, 371–87.

SAKTOR, B. (1953*b*). Investigations on the mitochondria of the house-fly, *Musca domestica* L. II. Oxidative enzymes with special reference to malic oxidase. *Arch. Biochem. Biophys.* **45**, 349–65.

SAKTOR, B. (1954). Investigations on the mitochondria of the house-fly, *Musca domestica* L. III. Requirements for oxidative phosphorylation. *J. gen. Physiol.* **37**, 343–59.

SAKTOR, B. (1955). Cell structure and the metabolism of insect flight muscle. *J. biophys. biochem. Cytol.* **1**, 1–12.

SAKTOR, B. and COCHRAN, D. (1956). Heterogeneity in the dinitrophenol uncoupling of mitochondrial oxidative phosphorylation. *J. Amer. chem. Soc.* **78**, 3227.

SAKTOR, B. and SANBORN, R. (1956). The effect of temperature on oxidative phosphorylation with insect flight muscle mitochondria. *J. biophys. biochem. Cytol.* **2**, 105–7.

SAKTOR, B., THOMAS, G. M., MOSER, J. C. and BLOCH, D. I. (1953). Dephosphorylation of adenosine triphosphate by tissues of the American cockroach, *Periplaneta americana* (L.). *Biol. Bull., Woods Hole,* **105**, 166–73.

SARA, M. and SMERDEL, A. (1953). Morfologia del dermascheletro del torace di *Asilus crabroniformis* L. (Dipt. Asilidae). *Ann. Ist. Mus. Zool. Univ. Napoli,* **5**, 1–31.

SARGENT, W. D. (1937). The internal thoracic skeleton of the dragonflies. *Ann. ent. Soc. Amer.* **30**, 81–95.

SCHELVER, F. J. (1802). Beobachtungen über der Flug und das Gesumme einiger zweiflügliger Insekten und insbesondere über die Schwingkölbchen und Schuppchen unter den Flügeldecken. *Wiedermanns Arch. Zool.* **2**, 210–18.

SCHNEIDER, G. (1953). Die Halteren der Schmeissfliege (*Calliphora*) als sinnesorgane und als mechanische flugstabilisatoren. *Z. vergl. Physiol.* **35**, 416–58.

SCHNEIDER, G. (1956). Zur spektralen Empfindlichkeit des Komplexauges von *Calliphora. Z. vergl. Physiol.* **39**, 1–20.

SELLKE, K. (1936). Biologische und morphologische studien an schädlichen Wiesenschnaken (Tipulidae, Dipt.). *Z. wiss. Zool.* **148**, 465–556.

SHEN, S. H. and YOUNG, B. (1943). Vision and flight; an experimental study on the cicada *Cryptotympana pustulata. Sinensia,* **14**, 55–60.

SIEBOLD, C. T. VON (1848). *Lehrbuch der vergleichenden Anatomie der wirbellosenen Thiere.* Berlin.

SLATER, E. C. and LEWIS, S. E. (1953). The effect of dinitrophenol on insect sarcosomes. *Biochem. J.* **55**, xxvii–xxviii.

SLIFER, E. H. and FINLAYSON, L. H. (1956). Muscle receptor organs in grasshoppers and locusts (Orthoptera, Acrididae). *Quart. J. micr. Sci.* **97**, 617–20.

SMART, J. (1957). The tergal depressor of the trochanter muscle in the Diptera. *Proc. Xth Int. Congr. Ent.*

SNODGRASS, R. E. (1909). The thorax of insects and the articulations of the wings. *Proc. U.S. Nat. Mus.* **36**, 511–95.

SNODGRASS, R. E. (1910). The thorax of the Hymenoptera. *Proc. U.S. Nat. Mus.* **39**, 37–91.

SNODGRASS, R. E. (1921). The seventeen-year locust. *Smithson. Rep., 1919,* 381–409.

SNODGRASS, R. E. (1925). *Anatomy and Physiology of the Honey-bee.* New York.

SNODGRASS, R. E. (1927). Morphology and mechanism of the insect thorax. *Smithson. misc. Coll.* **80**, no. 1.

SNODGRASS, R. E. (1929). The thoracic mechanism of a grasshopper, and its antecedents. *Smithson. misc. Coll.* **82**, no. 2.

127

SNODGRASS, R. E. (1935). *Principles of Insect Morphology.* New York.
SNODGRASS, R. E. (1942). The skeleto-muscular mechanisms of the honey-bee. *Smithson. misc. Coll.* **103**, no. 2, 1–120.
SNODGRASS, R. E. (1956). *The Anatomy of the Honey-bee.* Cornell University Press.
SOLF, V. (1931). Reizphysiologische Untersuchungen an Orthopteren-muskulatur. *Zool. Jb.*, Allg. Zool. Physiol., **50**, 175–264.
SOTAVALTA, O. (1947). The flight-tone (wing-stroke frequency) of insects. *Acta ent. fenn.* **4**, 1–117.
SOTAVALTA, O. (1952). The essential factor regulating the wing-stroke frequency of insects in wing mutilation and loading experiments and in experiments at subatmospheric pressure. *Ann. (bot.-zool.) Soc. zool.-bot. fenn. Vanamo (Zool.)*, **15**, 1–27.
SOTAVALTA, O. (1953). Recordings of high wing-stroke and thoracic vibration frequency in some midges. *Biol. Bull., Woods Hole*, **104**, 439–44.
SOTAVALTA, O. (1954a). On the thoracic temperature of insects in flight. *Ann. (bot.-zool.) Soc. zool.-bot. Vanamo (Zool.)*, **16**, no. 8.
SOTAVALTA, O. (1954b). The effect of wing inertia on the wing-stroke frequency of moths, dragonflies and cockroach. *Ann. ent. fenn.* **20**, 93–101.
STELLWAAG, F. (1910). Studien über die Honigbiene. II. Bau und Mechanik des Flugapparatus der Biene. *Z. wiss. Zool.* **95**, 518–550.
STELLWAAG, F. (1914). Der Flug der Lamellicornier. *Z. wiss. Zool.* **108**, 359–429.
STELLWAAG, F. (1916). Wie steuern die Insekten während des Fluges. *Biol. Zbl.* **36**, 30–44.
SUNDERMEIR, W. (1940). Der Hautpanzer des Kopfes und des Thorax von *Myrmeleon europaeus* und seine Metamorphose. *Zool. Jb.*, Anat u. Ontog. **66**, 291–348.
TAYLOR, G. I. (1952). Analysis of the swimming of long and narrow animals. *Proc. Roy. Soc.* A, **214**, 158–83.
THOM, A. and SWART, P. (1940). The forces on an aerofoil at very low speeds. *J. R. aero. Soc.* **44**, 761–70.
THOMAS, JOAN G. (1952). A comparison of the pterothoracic skeleton and flight muscles of male and female *Lamarckiana* species (Orthoptera, Acrididae). *Proc. R. ent. Soc. Lond.* A, **27**, 1–12.
THOMAS, JOAN G. (1953). A comparison of the flight muscles of Acrididae with different wing development. *Proc. R. ent. Soc. Lond.* A, **28**, 47–56.
TIEGS, O. W. (1955). The flight muscles of insects—their anatomy and histology; with some observations on the structure of striated muscle in general. *Phil. Trans.* B, **238**, 221–359.
VANDERPLANK, F. L. (1950). Air-speed/wing-tip speed ratios of insect flight. *Nature, Lond.*, **165**, 806–7.
VOGEL, R. (1911). Über die Innervierung der Schmetterlingsflügel und über den Bau und die Verbreitung der Sinnesorgane auf denselben. *Z. wiss. Zool.* **98**, 68–134.
VOGEL, R. (1912). Über die Chordotonalorgane in der Wurzel der Schmetterlingsflügel. *Z. wiss. Zool.* **100**, 210–44.

128

VOSKRESENKAYA, A. K. (1947). Functional peculiarities of the neuro-muscular apparatus of the wings of insects (in Russian). *J. Physiol. U.S.S.R.* **33**, 381–92.

VOSS, FR. (1905). Ueber den Thorax von *Gryllus domesticus*. *Z. wiss. Zool.* **78**, 268–354, 355–521, 645–96, 697–759.

VOSS, FR. (1913). Vergleichende Untersuchungen über die Flugwerkzeuge der Insekten. *Verh. dtsch. zool. Ges.* **23**, 118–42.

WALKER, G. T. (1925). The flapping flight of birds. *J. R. aero. Soc.* **29**, 590–4.

WALKER, G. T. (1927). The flapping flight of birds. II. *J. R. aero. Soc.* **31**, 337–42.

WATANABE, M. I. and WILLIAMS, C. M. (1951). Mitochondria in the flight muscles of insects. I. Chemical composition and enzymatic content. *J. gen. Physiol.* **34**, 675–89.

WEBER, H. (1927*a*). Die Gliederung der Sternalregion des Tenthrediniden-thorax. *Z. wiss. Insektenbiol.* **22**, 161–98.

WEBER, H. (1927*b*). Der Thorax der Hornisse. *Zool. Jb.*, Anat. u. Ontog. **47**, 1–100.

WEBER, H. (1928). Die Gliederung der Sternopleuralregion des Lepido-pterenthorax. *Z. wiss. Zool.* **131**, 181–254.

WEBER, H. (1929). Kopf und Thorax von *Psylla mali*. *Z. Morph. Öekol. Tiere,* **14**, 59–165.

WEBER, H. (1930). *Biologie der Hemipteren.* Berlin.

WEIS-FOGH, T. (1949). An aerodynamic sense-organ stimulating and regu-lating flight in locusts. *Nature, Lond.,* **164**, 873–4.

WEIS-FOGH, T. (1952). Fat combustion and metabolic rate of flying desert locusts (*Schistocerca gregaria* Forskål). *Phil. Trans.* B, **237**, 1–36.

WEIS-FOGH, T. (1956*a*). Biology and physics of locust flight. II. Flight performance of the desert locust (*Schistocerca gregaria*). *Phil. Trans.* B, **239**, 459–510.

WEIS-FOGH, T. (1956*b*). Biology and physics of locust flight. IV. Notes on sensory mechanisms in locust flight. *Phil. Trans.* B, **239**, 553–84.

WEIS-FOGH, T. (1956*c*). Tetanic force and shortening in locust flight muscle. *J. exp. Biol.* **33**, 668–84.

WEIS-FOGH, T. (1956*d*). The flight of locusts. *Sci. Amer.* March 1956.

WEIS-FOGH, T. and JENSEN, MARTIN (1956). Biology and physics of locust flight. I. Basic principles in insect flight. A critical review. *Phil. Trans.* B, **239**, 415–58.

WIERSMA, C. A. G., FURSHPAN, E. and FLOREY, E. (1953). Physiological and pharmacological observations on muscle receptor organs of the crayfish, *Cambarus clarkii* Girard. *J. exp. Biol.* **30**, 136–50.

WIGGLESWORTH, V. B. (1949). The utilization of reserve substances in *Drosophila* during flight. *J. exp. Biol.* **26**, 150–63.

WIGGLESWORTH, V. B. (1950). *The Principles of Insect Physiology.* London.

WILLIAMS, C. M., BARNES, L. A. and SAWYER, W. H. (1943). The utilization of glycogen by flies during flight and some aspects of the physiological ageing of *Drosophila*. *Biol. Bull.,* Woods Hole, **84**, 263–72.

WILLIAMS, C. M. and WILLIAMS, M. V. (1943). The flight muscles of *Drosophila repleta*. *J. Morph.* **72**, 589–97.

WITTIG, G. (1955). Untersuchungen am Thorax von *Perla abdominalis* Burm. (Larva und Imago). *Zool. Jb.*, Anat. u. Ontog. **74**, 491–570.

WOLF, E. and ZERRAHN-WOLF, G. (1936). The dark adaptation of the eye of the honey-bee. *J. gen. Physiol.* **19**, 229–38.

ZAĆWILICHOWSKI, J. (1933*a*). Über die Innervierung und die Sinnesorgane der Flügel von Schnabelfliegen (*Panorpa*). *Bull. int. Acad. Cracovie* (Acad. pol. Sci.), B, II, 109–124.

ZAĆWILICHOWSKI, J. (1933*b*). Über die Innervierung und die Sinnesorgane der Flügel der Honigbiene (*Apis mellifica* L.). *Bull. int. Acad. Cracovie* (Acad. pol. Sci.), B, II, 275–89.

ZAĆWILICHOWSKI, J. (1933*c*). Über die Innervierung und die Sinnesorgane der Flügel von Kocherfliegen (Trichoptera). *Bull. int. Acad. Cracovie* (Acad. pol. Sci.), B, II, 305–19.

ZAĆWILICHOWSKI, J. (1934*a*). Über die Innervierung und die Sinnesorgane der Flügel von Schabe *Phyllodromia germanica* L. *Bull. int. Acad. Cracovie* (Acad. pol. Sci.), B, II, 89–104.

ZAĆWILICHOWSKI, J. (1934*b*). Über die Innervierung und die Sinnesorgane der Flügel der Feldheuschrecke *Stauroderus biguttulus* (L.). *Bull. int. Acad. Cracovie* (Acad. pol. Sci.), B, II, 187–96.

ZAĆWILICHOWSKI, J. (1934*c*). Über die Innervierung und die Sinnesorgane der Flügel der Lausfliege *Oxypterum* Leach. (Diptera, Pupipara). *Bull. int. Acad. Cracovie* (Acad. pol. Sci.), B, II, 251–7.

ZAĆWILICHOWSKI, J. (1934*d*). Über die Innervierung und die Sinnesorgane des Flügels der Schnake *Tipula paludosa* Meig. *Bull. int. Acad. Cracovie* (Acad. pol. Sci.), B, II, 375–83.

ZAĆWILICHOWSKI, J. (1934*e*). Die Sinnesnervenelemente des Schwingers und dessen Homologie mit dem Flügel der *Tipula paludosa* Meig. *Bull. int. Acad. Cracovie* (Acad. pol. Sci.), B, II, 397–413.

ZAĆWILICHOWSKI, J. (1936*a*). Über die Innervation und die Sinnesorgane der Flügel von *Aphrophora alni* Fall. (Rhynchota-Homoptera). *Bull. int. Acad. Cracovie* (Acad. pol. Sci.), B, II, 85–99.

ZAĆWILICHOWSKI, J. (1936*b*). Über die Innervierung und Sinnesorgane der Flügel der Afterfrühlingsfliege *Isopteryx tripunctata* Scop. (Plecoptera). *Bull. int. Acad. Cracovie* (Acad. pol. Sci.), B, II, 267–84.

ZALOCKAR, M. (1947). Anatomie du Thorax de *Drosophila melanogaster*. *Rev. suisse. Zool.* **54**, 17–54.

ZEBE, E. (1957). Studies on glycolytic enzymes in the insect body. *Proc. Xth Int. Congr. Ent.*

INDEX OF INSECTS

INDEX OF SUBJECTS